Physiology at a Glance

Physiology at a Glance

JEREMY P.T. WARD

PhD
Professor of Respiratory Cell Physiology
School of Medicine
King's College London
London

ROBERT W. CLARKE

PhD
Senior Lecturer in Animal Physiology
Division of Animal Physiology
University of Nottingham
Loughborough

ROGER W.A. LINDEN

BDS PhD MFDS RCS
Professor of Craniofacial Biology and Head of Academic Department of Physiology
School of Biomedical Sciences
King's College London
London

Blackwell
Publishing

First published 2005

Library of Congress Cataloging-in-Publication Data
Ward, Jeremy P.T.
   Physiology at a glance / Jeremy Ward, Roger Linden, Rob Clarke.
      p. ; cm.
   Includes index.
   ISBN 1-4051-1328-6
   1. Physiology—Outlines, syllabi, etc.
   [DNLM: 1. Physiology—Handbooks.   QT 29 W259p 2005]   I. Linden, R. W. A.
(Roger W. A.)   II. Clarke, Rob.   III. Title.
   QP41.W268 2005
612—dc22
                                                            2004012015
   ISBN-13: 978-14051-1328-1
   ISBN-10: 1-4051-1328-6

Set in 9/11.5 Times by SNP Best-set Typesetter Ltd., Hong Kong
Printed and bound in India by Replika Press Pvt. Ltd

Commissioning Editor: Martin Sugden
Managing Editor: Geraldine Jeffers
Production Editor: Lorna Hind
Production Controller: Kate Charman
Text and cover designer: Simon Witter

For further information on Blackwell Publishing, visit our website:
http://www.blackwellpublishing.com

The publisher's policy is to use permanent paper from mills that operate a sustainable forestry
policy, and which has been manufactured from pulp processed using acid-free and elementary
chlorine-free practices. Furthermore, the publisher ensures that the text paper and cover board
used have met acceptable environmental accreditation standards.

Contents

Preface

Physiology is defined as 'The scientific study of the bodily function of living organisms and their parts'. There is a natural symbiosis between function (physiology) and structure (anatomy) from which physiology emerged as a separate discipline in the late part of the 19th century.

During the course of the 20th century reductionism took hold. Firstly biochemistry was born from the combination of chemistry with physiology, then soon after this there followed the disciplines of pharmacology, molecular biology and genetics. Now we are in the 21st century the discipline of physiology is again becoming important as it is realized that in order to make sense of developments such as the Human Genome Project we have to understand body function as an integrated whole. Thus, following a century of reductionism, we must attempt to put the whole back together again, and the study of whole-body physiology is about to be revitalized.

This volume is designed as a concise guide and revision aid to core topics in physiology, and should be useful to all students following a first year physiology course, whether they are studying single honours, biomedical sciences, nursing, medicine or dentistry. It should also be useful to those studying system-based curricula. The layout of *Physiology at a Glance* follows that of the other volumes in the *At a Glance* series, with a two-page spread for each topic (loosely corresponding to a lecture), comprising a large diagram on one page and concise explanatory text on the other.

Physiology is a large subject, and in a book this size we cannot hope to cover anything but the core and basics. As such, *Physiology at a Glance* should be used primarily to assist basic understanding and as an assistance to revision. Deeper knowledge should be gained by reference to full physiology and system textbooks and original peer-reviewed papers, especially in third year honours programmes. Students may find one or two sections of this book difficult, such as that on the physics of flow and diffusion, and such material may indeed not be included in some introductory Physiology courses. However, understanding these concepts often assists in learning how body systems behave in the way they do.

Whilst writing this book we have had immense help from several of our colleagues who have advised, and in some case chastised us on the content. We thank all those who have given us such advice; any errors are ours and not theirs. We would also like to thank the team at Blackwell Publishing, and in particular Geraldine Jeffers who provided great encouragement and support throughout the project, and Lorna Hind for all her work during the final production cycle.

Jeremy Ward
Roger Linden
Rob Clarke
August 2004

Robert Clarke (17th December 1956–25th August 2004)

On 25th August 2004, Rob Clarke died after a long, protracted illness. Rob was a dedicated researcher and teacher in physiology and it was typical of him to spend a considerable amount of the final year of his shortened life in producing this book. We were inspired by his determination to complete the book. Unfortunately he never saw the finished product and died two days before the final page proofs arrived on our desks.

We dedicate our contributions to Rob and share his love of physiology.

Roger Linden
Jeremy Ward
September 2004

Acknowledgements

Some figures in this book are taken from:
Ward, J.P.T. *et al.* (2002) *The Respiratory System at a Glance.* Blackwell Science, Oxford.
Aaronson, P. *et al.* (2003) *The Cardiovascular System at a Glance.* Blackwell Science, Oxford.

List of abbreviations

1,25-(OH)$_2$D	1,25-dihydroxycholecalciferol		GHRH	growth hormone releasing hormone
2,3-DPG	2,3-diphosphoglycerate		GIP	gastric inhibitory peptide
5-HT	5-hydroxytryptamine; serotonin		GLUT-1, 2 or 4	glucose transporters
ACE	angiotensin converting enzyme		GnRH	gonadotrophin releasing hormone
ACh	acetylcholine		GPCR	G-protein-coupled receptor
ACTH	adrenocorticotrophic hormone		G-Protein	GTP binding protein
ADH	antidiuretic hormone		GRP	gastrin-releasing peptide
ADP	adenosine diphosphate		GTP	guanosine triphosphate
AIDS	acquired immune deficiency syndrome		GTPase	enzyme that splits GTP
ANP	atrial natriuretic peptide		hCG	human chorionic gonadotrophin
ANS	autonomic nervous system		HIV	human immunodeficiency virus
AP	action potential		HMWK	high molecular weight kininogen
ATP	adenosine triphosphate		ICF	intracellular fluid
ATPase	enzyme that splits ATP		IG$_{A,E,G,M}$	immunoglobulin$_{A,E,G\ or\ M}$
AV node	atrio-ventricular node (heart)		IGF-1 or 2	insulin-like growth factor (1 or 2)
BTPS	body temperature and pressure, saturated with water		IL-1β or 6	interleukin-1β or 6
			IP$_3$	inositol trisphosphate
CAM-kinase	calcium-calmodulin kinase		IRS-1	insulin receptor substrate 1
cAMP	cyclic adenosine monophosphate		ISF	interstitial fluid
CCK	cholecystokinin		JAK	janus kinase
CDI	central diabetes insipidus		JGA	juxtaglomerular apparatus
cGMP	cyclic guanosine monophosphate		LH	luteinizing hormone
CICR	Ca^{2+}-induced Ca^{2+}-release		m or tRNA	messenger or transfer ribonucleic acid
COMT	catechol-O-methyl transferase		MAO	mono-amine oxidase
COX	cyclo-oxygenase		MAP	mean arterial pressure
CRH	corticotrophin releasing hormone		MAPK(K)	mitogen-activated protein kinase (kinase)
CVP	central venous pressure		MEPP	miniature end plate potentials
Da	dalton (molecular weight)		MIH	melanotrophin inhibiting hormone
DAG	diacylglycerol		MSH	melanotrophin stimulating hormone
DHEA	dehydroepiandrosterone		Na$^+$ pump	Na$^+$-K$^+$ ATPase
D$_{L}$O$_2$	O$_2$ diffusing capacity in lung; transfer factor		NAD$^+$ or (H)	nicotinic adenine dinucleotide (oxidized and reduced forms)
DNA	deoxyribonucleic acid			
DOPA	dihydroxyphenylalanine		NDI	nephrogenic diabetes insipidus
E$_{(Ion)}$	equilibrium potential for ion (K$^+$, Na$^+$, Ca^{2+} or Cl$^-$)		NGF	nerve growth factor
			NO	nitric oxide
ECF	extracellular fluid		NOS	nitric oxide synthase
ECG (EKG)	electrocardiogram (or graph)		PAH	*para*-aminohippuric acid
EDP	end diastolic pressure		PDGF	platelet-derived growth factor
EDV	end diastolic volume		PI-3 kinase	phosphatidylinositol-3 kinase
EGF	epidermal growth factor		pK	negative log of dissociation constant (buffers)
E$_m$	membrane potential			
EPO	erythropoietin		PKC	protein kinase C
EPP	end plate potential		PTH	parathyroid hormone
ESV	end systolic volume		Ras	a cellular GTPase
F$_{ab}$	hypervariable region of antibody molecule		ROC	receptor operated channels
F$_c$	constant region of antibody molecule		RTK	receptor tyrosine kinase
FEV$_1$	forced expiratory volume in 1 sec		SA node	sinoatrial node
FGF	fibroblast growth factor		SERCA	smooth endoplasmic reticulum Ca^{2+}-ATPase
*F*N$_2$	fractional concentration of nitrogen in a gas mixture		SMAD	intracellular protein associated with S-TKs
			SOC	store operated channels
FVC	forced vital capacity		SR	sarcoplasmic reticulum
FSH	follicle-stimulating hormone		SST	somatostatin
GDP	guanosine diphosphate		STAT	signal transduction and activation of transcription (protein)
GFR	glomerular filtration rate			
GH	growth hormone		S-TK	serine-threonine kinase

STPD	standard temperature and pressure, dry gas	TRα	thyroid hormone receptor
$T_{1, \text{or } 2}$	mono- or di-iodotyrosine	TRE	thyroid response element
T_3	tri-iodothyronine	TSH	thyroid-stimulating hormone
T_4	thyroxine	TV	tidal volume
TGFβ	transforming growth factor β	TXA_2	thromboxane A_2
TH	thyroid hormone	UCP-1, 2 or 3	uncoupling protein-1, 2 or 3
T_m	tubular transport maximum (kidney)	V_A/Q mismatch	ventilation-perfusion mismatch (lungs)
TNF	tumour necrosis factor	VIP	vasoactive intestinal polypeptide

1 Physiology and the genome

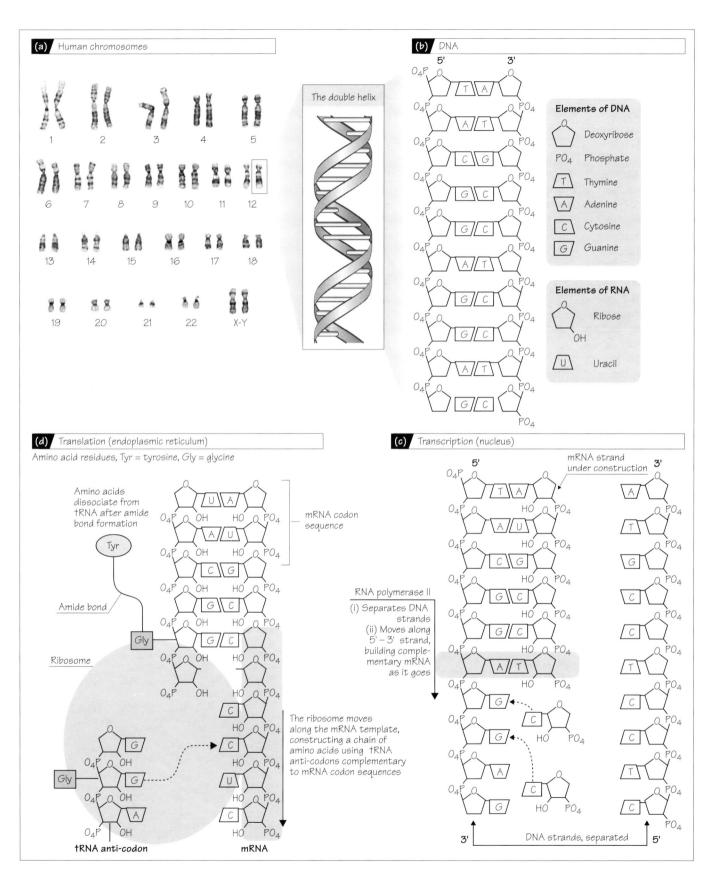

(a) Human chromosomes

1 2 3 4 5

6 7 8 9 10 11 12

13 14 15 16 17 18

19 20 21 22 X-Y

The double helix

(b) DNA

5'   3'

Elements of DNA

Deoxyribose
PO₄   Phosphate
T   Thymine
A   Adenine
C   Cytosine
G   Guanine

Elements of RNA

Ribose
OH
U   Uracil

(d) Translation (endoplasmic reticulum)

Amino acid residues, Tyr = tyrosine, Gly = glycine

Amino acids dissociate from tRNA after amide bond formation

Tyr

Amide bond

Gly

Ribosome

mRNA codon sequence

The ribosome moves along the mRNA template, constructing a chain of amino acids using tRNA anti-codons complementary to mRNA codon sequences

tRNA anti-codon   mRNA

(c) Transcription (nucleus)

5'   mRNA strand under construction   3'

RNA polymerase II

(i) Separates DNA strands
(ii) Moves along 5' – 3' strand, building complementary mRNA as it goes

DNA strands, separated

3'   5'

DNA and genes

The genetic heritage of *homo sapiens* amounts to some 30 000 genes packed into 44 somatic chromosomes arranged in pairs, plus the sex chromosomes X and Y (Fig. 1a). Multicellular organisms, such as ourselves, are made up of **eukaryotic** cells, which means that the genes are stored in the cell **nucleus** (Chapter 4). The genetic material comprises **deoxyribonucleic acid** (**DNA**; Fig. 1b), an alternating chain of phosphate and sugar (deoxyribose) residues, with **nucleotide bases** attached to each sugar moiety. This molecule self-organizes into a helix made up of two polarized, complementary strands that are arranged so that the phosphate terminal (**5′ terminus**) of one is opposite the sugar terminal (**3′ terminus**) of the other.

The two filaments of the helix are held together by hydrogen bonds between pairs of bases. The nucleotides found in DNA are **adenine** (**A**), **cytosine** (**C**), **guanine** (**G**) and **thymine** (**T**). An adenine on one strand of the helix will always be found opposite a thymine residue on the other, whereas cytosine is always found opposite guanine. The human genome comprises some 3.2×10^9 (3.2 billion) of these **base pairs**. The linear arrangement of bases forms the genetic code or **genome**. The working draft of the human genome that was announced in 2001 describes the sequences of base pairs in each chromosome. This information alone does not tell us where all the genes are located, as only some 1.5% of the genome is accounted for by translatable code. Each gene codes for a specific protein, and a large number of gene sequences have been identified by the process of **cloning**. The average length of these sequences in humans is roughly 27 000 base pairs, even though an average-sized protein requires a code of not much more than 1000 base pairs. The useful DNA code is contained within short sequences, known as **exons**, but these are intermingled with much longer sections, known as **introns**, that do not code for protein. In addition, each gene has a number of regulatory sequences that allow the gene to be switched on and off and indicate where the gene begins and ends. Not only does each gene contain large sections of non-coding DNA, but between working genes there are interposed long sequences of what appears to be non-useful, so-called 'mobile' material that has the ability to replicate itself for insertion at points along the DNA molecule.

Gene transcription and translation

The decoding of a gene into a protein is a two-stage process involving **transcription** of DNA into **messenger ribonucleic acid** (**mRNA**), followed by **translation** of mRNA into protein. In eukaryotic organisms, the DNA code is read by an enzyme known as **RNA polymerase II** which constructs sequences of mRNA from the DNA template. RNA polymerase II attaches to DNA with the help of a set of proteins, known as **general transcription factors**. These factors identify the starting sequence of a gene (the **promoter** region), unravel the chromosomal material (within which DNA is very tightly coiled) and unzip a portion of the paired helix to allow the polymerase to bind. Only one strand of the DNA is read at any one time, and transcription always begins at the 5′ end of a sequence. RNA polymerase II moves along the gene sequence base by base, creating an mRNA strand that is complementary to the original DNA (Fig. 1c). In RNA, the sugar residue in the backbone is ribose, and the base **uracil** (**U**) substitutes for thymine. Thus, mRNA is constructed with a uracil for each adenine on the DNA, a cytosine for a guanine, and so on, until the sequence indicating the end of the gene is reached. During the process, any RNA derived from introns is removed and transcription errors are corrected. Once complete, the mRNA strand moves from the nucleus to the **endoplasmic reticulum** (Chapter 4) where structures known as **ribosomes** convert the RNA message into protein. Within RNA, strings of three consecutive bases (**codons**) code for one of the 20 amino acids used in human proteins, or act as start or stop signals. **Transfer RNA** (**tRNA**) exists in three-base sequences, known as **anticodons**, which are complementary to the codons present in the mRNA (Fig. 1d). Each type of tRNA binds a specific amino acid so that, as the ribosome moves along the mRNA strand, a chain of amino acids is formed that mirrors the codon sequence of the RNA message.

What the genome tells us

The protein products of gene transcription organize themselves and other biological molecules into living organisms. The functions of these proteins and the assemblies they produce are at least as important as the genome itself in determining the nature and operational characteristics of organisms. The mouse genome is more than 90% homologous to that of humans, yet there are some striking differences between the two species, some of which must be explained by postgenomic processes. Although the sequencing of the genome is a great achievement, in isolation it tells us nothing about the functioning of the human organism. Such insight can only come from investigations on higher level systems, taking advantage of the new opportunities provided by our knowledge of the code. This is modern physiology.

2 Homeostasis and the physiology of proteins

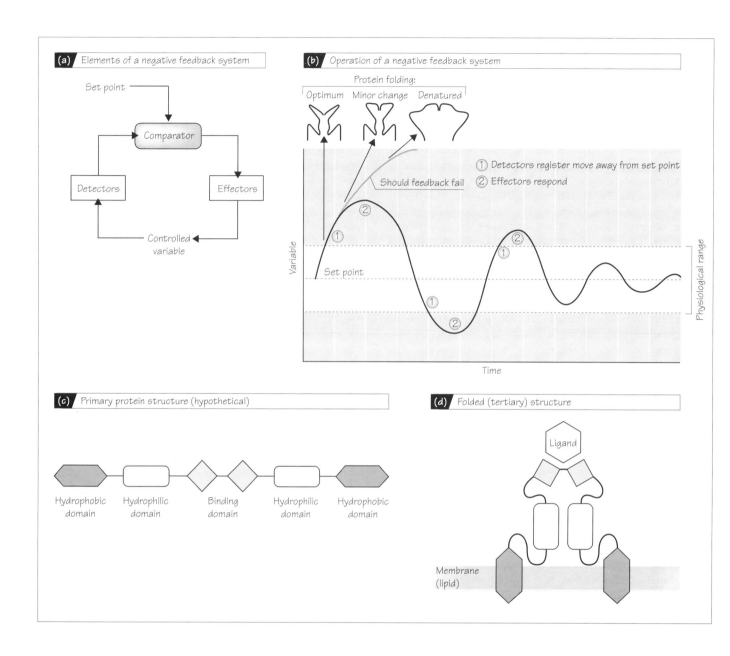

(a) Elements of a negative feedback system

Set point

Comparator

Detectors

Effectors

Controlled variable

(b) Operation of a negative feedback system

Protein folding:

Optimum   Minor change   Denatured

① Detectors register move away from set point
② Effectors respond

Should feedback fail

Variable

Set point

Time

Physiological range

(c) Primary protein structure (hypothetical)

Hydrophobic domain   Hydrophilic domain   Binding domain   Hydrophilic domain   Hydrophobic domain

(d) Folded (tertiary) structure

Ligand

Membrane (lipid)

Homeostasis is the ability of physiological systems to maintain conditions within the body in a relatively constant state. It is probably the most important concept in physiology.

Negative feedback control

Homeostatic mechanisms provide the tight regulation of *all* physiological variables and work on the principle of **negative feedback**. A negative feedback system (Fig. 2a) comprises: **detectors** (often neural **receptor cells**) to measure the variable in question; a **comparator** (usually a neural assembly in the central nervous system) to receive input from the detectors and compare the size of the signal against the desired level of the variable (the **set point**); and **effectors** (muscular and/or glandular tissue) that are activated by the comparator to restore the variable to its set point. The term 'negative feedback' comes from the fact that the effectors always act to move the variable in the opposite direction to the change that was originally detected. Thus, when the partial pressure of CO_2 in blood increases above 40 mmHg, brain stem mechanisms increase the rate of ventilation to clear the excess gas, and vice versa when CO_2 levels fall (Chapter 25). The term 'set point' implies that there is a single optimum value for each physiological variable; however, there is some tolerance in all physiological systems and the set point is actually a narrow *range* of values within which physiological processes will work normally (Fig. 2b). Not only is the set point not a point, but it can be reset in some systems according to physiological requirements. For instance, at high altitude, the low partial pressure of O_2 in inspired air causes the ventilation rate to increase. Initially, this effect is limited due to the loss of CO_2, but, after 2–3 days, the brain stem lowers the set point for CO_2 control and allows ventilation to increase further, a process known as **acclimatization**.

A common operational feature of all negative feedback systems is that they induce oscillations in the variable that they control (Fig. 2b). The reason for this is that it takes time for a system to detect and respond to a change in a variable. This delay means that feedback control always causes the variable to overshoot the set point slightly, activating the opposite restorative mechanism to induce a smaller overshoot in that direction, until the oscillations fall within the range of values that are optimal for physiological function. Normally, such oscillations have little visible effect. However, if unusually long delays are introduced into a system, the oscillations can become extreme. Patients with congestive heart failure sometimes show a condition known as **Cheyne–Stokes' breathing**, in which the patient undergoes periods of deep breathing interspersed with periods of no breathing at all (**apnoea**). This is partly due to the slow flow of blood from the lungs to the brain, which causes a large delay in the detection of blood levels of CO_2.

Some physiological responses use *positive* feedback, causing rapid amplification. Examples include initiation of an action potential, where sodium entry causes depolarization which therefore increases sodium entry and thus more depolarization (Chapter 6), and certain hormonal changes particularly in reproduction (Chapter 48). Positive feedback is inherently unstable, and requires some mechanism to break the feedback loop and stop the process, such as time-dependent inactivation of sodium channels for the first example.

Protein form and function are protected by homeostatic mechanisms

The homeostatic mechanisms that are described in detail throughout this book have evolved to protect the integrity of the protein products of gene translation. Normal functioning of proteins is essential for life, and usually requires binding to other molecules, including other proteins. The specificity of this binding is determined by the three-dimensional shape of the protein. The **primary structure** of a protein is determined by the sequence of amino acids, and genetic mutations that alter this sequence can have profound effects on the functionality of the final molecule. Such gene **polymorphisms** are the basis of many genetically based disorders. The final shape of the molecule (the **tertiary structure**), however, results from a process of **folding** of the amino acid chain (Fig. 2d). Folding is a complex process by which a protein achieves its lowest energy conformation. It is determined by electrochemical interactions between amino acid side-chains (e.g. hydrogen bonds, van der Waals' forces), and is so vital that it is overseen by **molecular chaperones**, such as the **heat shock proteins**, which provide a quiet space within which the protein acquires its final shape. In healthy tissue, cells can detect and destroy misfolded proteins, the accumulation of which damages cells and is responsible for various pathological conditions, including **Alzheimer's disease** and **Creutzfeldt–Jakob disease**. Folding ensures that functional sequences of amino acids (**domains**) that form, for instance, binding sites for other molecules or hydrophobic segments for insertion into a membrane are properly orientated to allow the protein to serve its function.

The relatively weak nature of the forces that cause folding renders them sensitive to changes in the environment surrounding the protein. Thus, alterations in acidity, osmotic potential, concentrations of specific molecules/ions, temperature or even hydrostatic pressure can modify the tertiary shape of a protein and change its interactions with other molecules. These modifications are usually reversible and are exploited by some proteins to detect alterations in the internal or external environments. For instance, nerve cells that respond to changes in CO_2 (chemoreceptors; Chapter 25) possess **ion channel** proteins (Chapter 5) that open or close to generate electrical signals (Chapter 6) when the acidity of the medium surrounding the receptor alters by more than a certain amount (CO_2 forms an acid in solution). However, there are limits to the degree of fluctuation in the internal environment that can be tolerated by proteins before their shape alters so much that they become non-functional or irreversibly damaged, a process known as **denaturing** (this is what happens to egg white proteins in cooking). Homeostatic systems prevent such conditions from arising within the body, and thus preserve protein functionality.

3 Body water compartments and physiological fluids

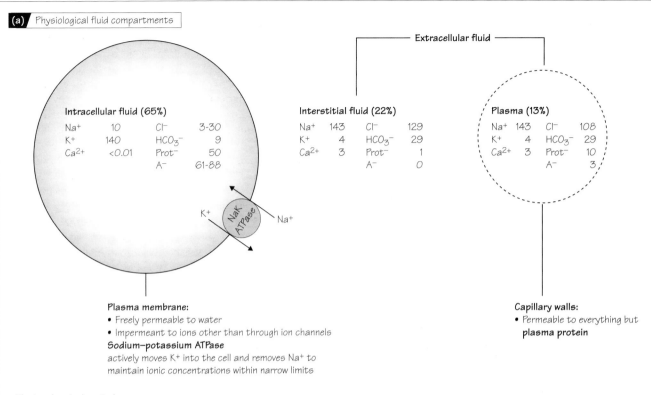

(a) Physiological fluid compartments

Extracellular fluid

Intracellular fluid (65%)

Na^+	10	Cl^-	3-30
K^+	140	HCO_3^-	9
Ca^{2+}	<0.01	$Prot^-$	50
		A^-	61-88

Interstitial fluid (22%)

Na^+	143	Cl^-	129
K^+	4	HCO_3^-	29
Ca^{2+}	3	$Prot^-$	1
		A^-	0

Plasma (13%)

Na^+	143	Cl^-	108
K^+	4	HCO_3^-	29
Ca^{2+}	3	$Prot^-$	10
		A^-	3

K^+ NaK ATPase Na^+

Plasma membrane:
- Freely permeable to water
- Impermeant to ions other than through ion channels

Sodium–potassium ATPase
actively moves K^+ into the cell and removes Na^+ to maintain ionic concentrations within narrow limits

Capillary walls:
- Permeable to everything but **plasma protein**

Electrochemical equivalence:
Figures are given in $mEq\ L^{-1}$. An equivalent (Eq) is 1 mole of charge, so that 1 Eq of a monovalent ion such as Na^+ is 1 mole, but 1 Eq of Ca^{2+} is 0.5 mole. $Prot^-$ and A^- represent the fixed intracellular anions: A^- is mainly PO_4^{3-} but includes SO_4^{3-}.

Compartment volumes:
In a 70 kg person, the approximate volumes of each compartment would be:
Intracellular fluid 27 L, interstitial fluid 9.5 L, plasma 3.5 L.

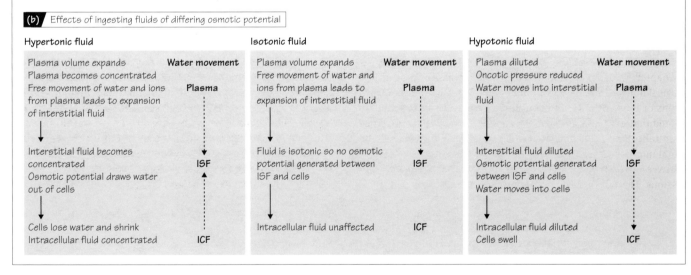

(b) Effects of ingesting fluids of differing osmotic potential

Hypertonic fluid

	Water movement
Plasma volume expands	
Plasma becomes concentrated	
Free movement of water and ions from plasma leads to expansion of interstitial fluid	Plasma
↓	↓
Interstitial fluid becomes concentrated	
Osmotic potential draws water out of cells	ISF ↑
↓	
Cells lose water and shrink	
Intracellular fluid concentrated	ICF

Isotonic fluid

	Water movement
Plasma volume expands	
Free movement of water and ions from plasma leads to expansion of interstitial fluid	Plasma
↓	↓
Fluid is isotonic so no osmotic potential generated between ISF and cells	ISF ↑
↓	
Intracellular fluid unaffected	ICF

Hypotonic fluid

	Water movement
Plasma diluted	
Oncotic pressure reduced	
Water moves into interstitial fluid	Plasma
↓	↓
Interstitial fluid diluted	
Osmotic potential generated between ISF and cells	ISF
Water moves into cells	
↓	
Intracellular fluid diluted	
Cells swell	ICF ↓

Osmosis

Osmosis is the passive movement of water across a **semipermeable membrane** from regions of low solute concentration to those of higher solute concentration. Biological membranes are semipermeable in that they usually allow the free movement of water but restrict the movement of solutes. The creation of **osmotic gradients** is the primary method for the movement of water in biological systems. This is why the osmotic potential (**osmolality**) of body fluids is closely regulated by a number of homeostatic mechanisms (Chapter 31). A fluid at the same osmotic potential as plasma is said to be **isotonic**; one at higher potential (i.e. more concentrated solutes) is **hypertonic** and one at lower potential is **hypotonic**. Drinking fluids of differing osmotic potentials has distinct effects on the distribution of water between cells and extracellular fluid (Fig. 3b). An osmole is one mole of osmotically active particles irrespective of their identity. The osmotic potential is expressed as the **osmolarity**, measured in $osmol\,L^{-1}$, or the **osmolality**, measured in $osmol\,(kg\,H_2O)^{-1}$. The latter is preferred by physiologists as it is independent of temperature.

Body water compartments

Water is the solvent in which almost all biological reactions take place (the other being membrane lipid), and so it is fitting that it accounts for some 50–70% of the body mass (i.e. about 40 L for a 70 kg person). The nature of biological membranes means that water moves freely within the body, but the materials dissolved in it do not. There are two major 'fluid compartments': the water within cells (**intracellular fluid, ICF**), which accounts for about 65% of the body total, and the water outside cells (**extracellular fluid, ECF**). These compartments are separated by the plasma membranes of the cells, and differ markedly in terms of the concentrations of the ions that are dissolved in them (Fig. 3a; Chapter 5). Approximately 65% of the ECF comprises the tissue fluid found between cells (**interstitial fluid, ISF**), and the rest is made up of the liquid component of blood (**plasma**). The barrier between these two fluids consists of the walls of tiny blood vessels, known as capillaries (Fig. 3a; Chapter 19).

Intracellular vs. extracellular fluid

Many critical biological events, including all bioelectrical signals (Chapter 6), depend on maintaining the composition of physiological fluids within narrow limits. Figure 3a shows the concentrations of ions in the three main fluid compartments. It should be noted that, *within* any one compartment, there *must* be electrical neutrality, i.e. the total number of positive charges must equal the total number of negative charges. The most important difference between ICF and ECF lies in the relative concentrations of cations. The K^+ ion concentration is much higher inside the cell than in ECF, whilst the opposite is true for the Na^+ ion concentration. Ca^{2+} and Cl^- ion concentrations are also higher in ECF. The question arises as to how these differences come about, and how they are maintained. Ion channel proteins allow the cell to determine the flow of ions across its own membrane (Chapter 5). In most circumstances, relatively few channels are open so that the leakage of ions is low. There is, however, always a steady movement of ions across the membrane, with Na^+ and K^+ following their concentration gradients into and out of the cell, respectively. Uncorrected, the leak would eventually lead to the equalization of the compositions of the two compartments, effectively eliminating all bioelectrical signalling (Chapter 6). This is prevented by the activity of the sodium–potassium ATPase (Chapter 4). Of the other ions, most Ca^{2+} in the cell is transported actively into the endoplasmic reticulum and mitochondria, leaving very low levels of free Ca^{2+} in ICF. Cl^- ions are differentially distributed across the membrane by virtue of their negative charge. Intracellular proteins (which are numerous; Chapter 4) are negatively charged at physiological pH. These, and other large anions that cannot cross the plasma membrane (e.g. phosphate, PO_4^{3-}), are trapped within the cell and account for most of the anion content of ICF. Cl^- ions, which *can* diffuse across the membrane through channels, are forced out of the cell by the charge on the fixed anions. The electrical force driving Cl^- ions out of the cell is balanced by the chemical gradient driving them back in, a situation known as the **Donnan equilibrium**. Variations in the large anion content of cells mean that the concentration of Cl^- ions in ICF can vary by a factor of 10 between cell types, being as high as 30 mM in cardiac myocytes, although lower values (around 5 mM) are more common.

Interstitial fluid vs. plasma

The main difference between these fluids is that plasma contains more protein than does ISF (Fig. 3a). The plasma proteins (Chapter 9) are the only constituents of plasma that do not cross into ISF, although they are allowed to escape from capillaries in very specific circumstances (Chapter 10). The presence of impermeant proteins in the plasma exerts an osmotic force relative to ISF (**plasma oncotic pressure**) that almost balances the hydrostatic pressure imposed on the plasma by the action of the heart, which tends to force water out of the capillaries, so that there is a small net water movement out of the plasma into the interstitial space. The leakage is absorbed by the **lymphatic system** (Chapter 19). **Transcellular fluid** is the name given to fluids that do not contribute to any of the main compartments, but which are derived from them. It includes cerebrospinal fluid and exocrine secretions, particularly gastrointestinal secretions (Chapters 33–37), and has a collective volume of approximately 2 L.

4 Cells, membranes and organelles

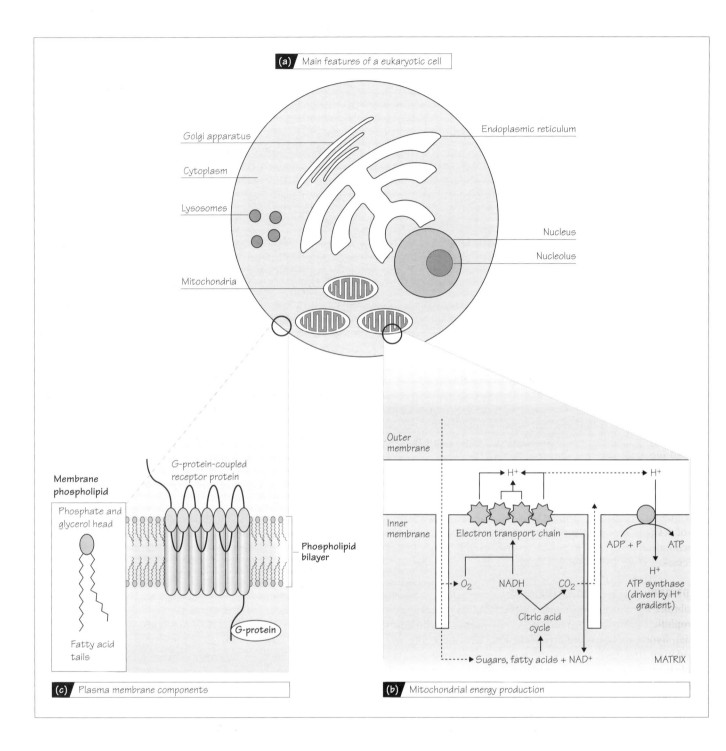

(a) Main features of a eukaryotic cell

Golgi apparatus

Cytoplasm

Lysosomes

Mitochondria

Endoplasmic reticulum

Nucleus

Nucleolus

Membrane phospholipid

Phosphate and glycerol head

Fatty acid tails

G-protein-coupled receptor protein

Phospholipid bilayer

G-protein

(c) Plasma membrane components

Outer membrane

Inner membrane

H^+

H^+

Electron transport chain

O_2 NADH CO_2

Citric acid cycle

ADP + P ATP

H^+

ATP synthase (driven by H^+ gradient)

Sugars, fatty acids + NAD^+

MATRIX

(b) Mitochondrial energy production

The aqueous internal environment of the cell is separated from the aqueous external medium by an envelope of fat molecules (**lipids**) known as the **plasma membrane**, which is described in detail below.

Protein-processing organelles

Just over half the content of the cell is made up of **cytosol**, a viscous, protein-rich fluid that fills the spaces between the internal structures of the cell. The remainder of the cell is made up of a number of subcellular systems or **organelles** that are themselves enclosed by lipid membranes. The most obvious of these is the **nucleus** (Fig. 4.1a), which contains the chromosomes, and the **nucleolus**, a structure that lacks a surrounding membrane and which is responsible for the assembly of ribosomes (Chapter 1). The **endoplasmic reticulum** is the site of protein assembly by ribosomes and, in concert with the **Golgi apparatus**, performs post-translational processing of newly made proteins. This includes trimming the amino acid chain to the right length, the control of folding, the addition of polysaccharide chains to strategic positions in the molecule (**glycosylation**) and the identification of improperly folded proteins, which are tagged for subsequent destruction. The endoplasmic reticulum also serves as a store for intracellular Ca^{2+} ions and is the major site of lipid production in the cell. Fully processed proteins are delivered from the Golgi apparatus to various intracellular destinations. Thus, receptor, transport and structural proteins are sent to the plasma membrane, digestive enzymes go to the lysosomes and signal molecules and enzymes for extracellular action are packaged into secretory vesicles. **Lysosomes** contain several acid hydrolase enzymes collectively capable of catabolizing any cellular macromolecule, that work optimally at pH 5.0. The low operating pH means that any lysosomal enzymes leaking into the cytosol will not attack the host cell as long as the cellular pH is near the physiological value of 7.2. Lysosomes digest unwanted materials, including improperly folded proteins, thus recycling raw materials and preventing the accumulation of potentially problematic rubbish within the cell (Chapter 2).

Mitochondria and energy production

Mitochondria use molecular oxygen to, in effect, burn sugar and small fatty acid molecules to produce **adenosine triphosphate** (**ATP**), the compound that is used to drive all energy-requiring reactions in cells. Sugars are first converted to pyruvate in the cytosol by glycolysis. The outer membrane of the mitochondrion is freely permeable to all molecules in the cytosol, but the highly folded inner membrane allows only small fuel molecules to pass into the **matrix** that it contains. Enzymes within the matrix drive the **citric acid cycle** (Fig. 4b) to turn pyruvate and fatty acids into reduced nicotinic adenine dinucleotide (**NADH**) and the waste product CO_2. The **electron transport chain**, a series of enzymes in the inner membrane, then uses molecular oxygen to re-oxidize NADH to NAD^+ and generate a hydrogen ion gradient across the inner membrane. This gradient is then used to drive the production of ATP by the enzyme **ATP synthase** (Fig. 4b). In this way, mitochondria provide the power for almost all physiological processes.

Membranes and membrane proteins

Membrane lipids (most of which are phospholipids) comprise a **hydrophilic** (water-loving) head, with two short (14–24 carbon atoms), **hydrophobic** (water-repelling) fatty acid tails (Fig. 4c). Placed in an aqueous medium, such molecules self-organize into a double layer (**bilayer**) with the heads facing outwards and the tails inwards (Fig. 4c). Lipid molecules diffuse freely within each layer (**lateral diffusion**) and the membrane is a fluid structure, with the lipid molecules continuously jostling each other from side to side. However, it is rare for lipid molecules to flip from one layer to the other. Non-polar, lipid-soluble molecules, such as O_2, CO_2 and small polar substances, such as water or urea, readily pass through the lipid bilayer. However, large polar molecules, such as glucose and ions of any type, cannot cross a purely lipid barrier at all. **Membrane proteins** provide the mechanisms that allow the transmembrane movement of such materials. Some proteins are attached to only one surface of the bilayer (intracellular or extracellular), whereas others pass right through the membrane.

Structural proteins, such as **spectrin**, are associated with the inner lipid layer, supplying a flexible supporting scaffold for the cell, known as the **cytoskeleton**. The cytoskeleton gives mechanical strength to the cellular envelope and allows it to withstand distortion. Other important intracellular membrane proteins are associated with cell signalling. For instance, **GTP-binding proteins** (**G-proteins**) are enzymes that cleave guanosine triphosphate (GTP) to guanosine diphosphate (GDP) to activate or inhibit other membrane-bound signalling enzymes, such as **adenylyl cyclase**, which generates cyclic adenosine monophosphate (**cAMP**) from ATP. cAMP activates **protein kinase** enzymes within the cell to initiate numerous changes in cell function by **phosphorylating** intracellular proteins. **Transmembrane** proteins (Fig. 4c) penetrate the entire thickness of the bilayer. The intramembrane segments are composed of chains of hydrophobic amino acid residues and the extra- and intracellular portions are made up predominantly of hydrophilic residues. Many such proteins belong to a super family of proteins, known as the **G-protein-coupled receptors** (GPCRs). These proteins possess seven membrane-spanning segments and detect signal molecules (neurotransmitters or hormones) in the extracellular medium (Fig. 4c). On binding the appropriate molecule, these GPCRs activate one of the G-proteins (Table 4) to initiate a biochemical cascade, resulting in altered cellular activity. Like lipid molecules, membrane proteins undergo lateral diffusion and move around the membrane. Furthermore, the cell can control exactly which proteins insert into which portion of the membrane. For instance, the cells that line the kidney tubules are polarized so that the **sodium–potassium ATPase** transporter proteins (Chapters 5 & 29) are located only on one side of the cell.

Table 4 The main guanosine triphosphate-binding proteins (G-proteins) and their intracellular actions

G-protein	Effect	Intracellular products	Intracellular effector
G_s	Activates adenylyl cyclase	↑ cAMP	Protein kinase A
G_i	Inhibits adenyly cyclase	↓ cAMP	Protein kinase A
G_q	Activates phospholipase C	↑ Diacylglycerol	Protein kinase C
		↑ Inositol triphosphate	Release of stored Ca^{2+}

cAMP, cyclic adenosine monophosphate.

5 Membrane transport proteins and ion channels

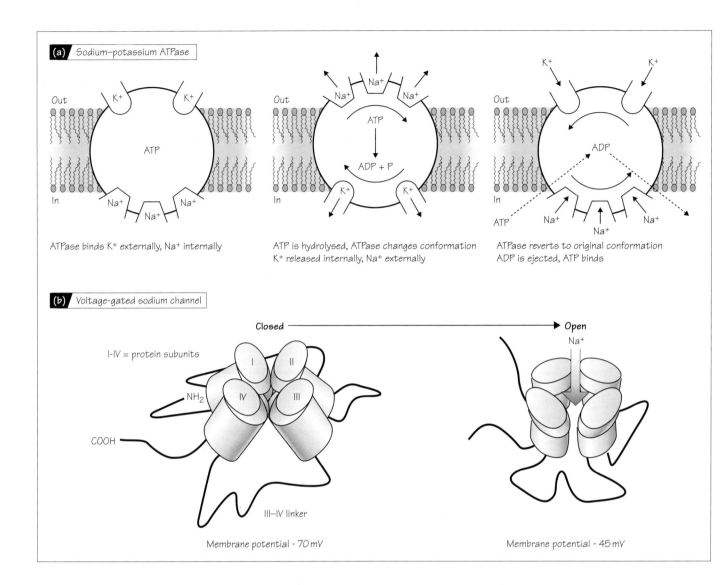

(a) Sodium–potassium ATPase

ATPase binds K+ externally, Na+ internally

ATP is hydrolysed, ATPase changes conformation
K+ released internally, Na+ externally

ATPase reverts to original conformation
ADP is ejected, ATP binds

Out

In

ATP

ADP + P

ADP

ATP

(b) Voltage-gated sodium channel

I-IV = protein subunits

Closed ──────────────────→ Open

Na+

NH₂

COOH

I II IV III

III–IV linker

Membrane potential - 70 mV

Membrane potential - 45 mV

Proteins provide several routes for the movement of materials across membranes: (i) large pores, constructed of several protein subunits, that allow the bulk flow of water, ions and sometimes larger molecules (e.g. **aquaporin**, Chapter 30, and the **connexins**, that form **gap junctions** between cells); (ii) transporter molecules, some of which use metabolic energy (either direct or indirect) to move molecules against chemical and/or electrical gradients; and (iii) ion channels, specialized to allow the passage of particular ion species across the membrane under defined conditions.

Carrier-mediated transport

Transporter (or carrier) proteins can move a single type of molecule in one direction across a membrane (a **uniporter**), several different molecules in one direction (a **symporter**) or different molecules in opposite directions (an **antiporter**). All types of transporter work on the same basic principle (Fig. 5a). Transporters can allow the movement of molecules down chemical concentration gradients (**facilitated diffusion**), when the energy required for conformational changes in the transporter protein is provided by the concentration gradient rather than by metabolic activity. Important transporters for glucose and amino acids, found in the kidney and the gut, are in fact driven by the Na^+ concentration gradient that exists across the cell membrane (Chapter 3). These symporters must bind sodium *and* the primary transported molecule at the external surface of the membrane before the conformational change will take place. This is known as **secondary active transport**, as the Na^+ gradient is set up by a process requiring metabolic energy. The uneven distribution of Na^+ ions across the cell membrane is produced by the best known of all transporters, the **sodium–potassium ATPase**, also known as the **sodium pump** (Fig. 5a). This protein is an antiporter that uses metabolic energy to move Na^+ ions out of the cell and K^+ ions in, *against their respective concentration gradients*. The ATPase binds extracellular K^+ and intracellular Na^+ ions, usually in the ratio of $2:3$, and hydrolyses adenosine triphosphate (ATP) to change its conformation, leading to the ejection of Na^+ into the extracellular medium and K^+ into the cytosol; this gives a high concentration of K^+ ions and a low concentration of Na^+ ions inside the cell (Chapter 3). The pump works continuously as long as the cell is alive, although its activity is stimulated by high intracellular levels of Na^+ ions. The action of the sodium–potassium ATPase is an example of **primary active transport**.

Ion channels

Ions can *diffuse* across cell membranes through **ion channels**. These transmembrane proteins, which are invariably constructed of several subunits, provide a charged, hydrophilic pore through which ions can move across the lipid bilayer. They possess a number of important features that confer upon the cell the ability to control closely the movement of all mobile ions. Ion channels are **selective** for particular ions, i.e. they allow the passage of only one type of ion or a few related ions. There are specialized channels for Na^+, K^+, Cl^- and Ca^{2+} ions, as well as non-specific channels for monovalent, divalent or even all types of cations (positively charged ions). The charge on the transmembrane pore determines whether the channel is for cations or anions (negatively charged ions), and selection between different ion types seems to depend on the size of the ion *without* its accompanying water of hydration. The second key feature of ion channels is that they can be opened or closed, a process known as **gating**. Gating is brought about by a change in the conformation of the protein subunits that opens or closes the ion-permeable pore (e.g. Fig. 5b). Many channels are opened or closed according to the electrical potential difference (voltage) across the cell membrane (**voltage** gating; Chapter 6), whereas others are gated by the presence of a specific signal molecule (**ligand** or **recptor** gating). Some ligand-gated channels are receptive to extracellular molecules, such as neurotransmitters or hormones, whereas others respond to intracellular signals, such as cyclic adenosine monophosphate (cAMP) (Chapter 4). As mentioned in Chapter 2, some cells are specialized to detect changes in the internal and external environments (receptor cells), and these possess ion channels that are gated by the particular signal that is detected by the receptor, e.g. pH or light. The characteristics of ion channels, in concert with the activities of ion pumps, give cells the ability to control precisely the movement of ions across the cell membrane. This is crucial for many important physiological processes, including electrical signalling (Chapters 6 & 7), initiation of muscle contraction (Chapters 50 & 51) and the release of materials such as neurotransmitters, hormones and digestive enzymes.

A note about diffusion

Diffusion, the passive movement of solutes down electrochemical gradients, is an important process in physiology. However, it is only really useful for processes that occur on a cellular scale. This is because the time taken for a molecule to diffuse a given distance is proportional to the square of that distance. Thus, over spaces of $1\,\mu m$ or smaller, diffusion is incredibly quick, with the ground covered in a few milliseconds or less. Up to $10\,\mu m$, the process is still reasonably rapid, occurring in seconds, but at $1\,cm$ and above the time taken for diffusion to occur starts to grow into minutes, which is insufficiently rapid for physiological processes. Factors that influence the rate of diffusion across biological membranes are discussed in Chapter 11.

6 Biological electricity

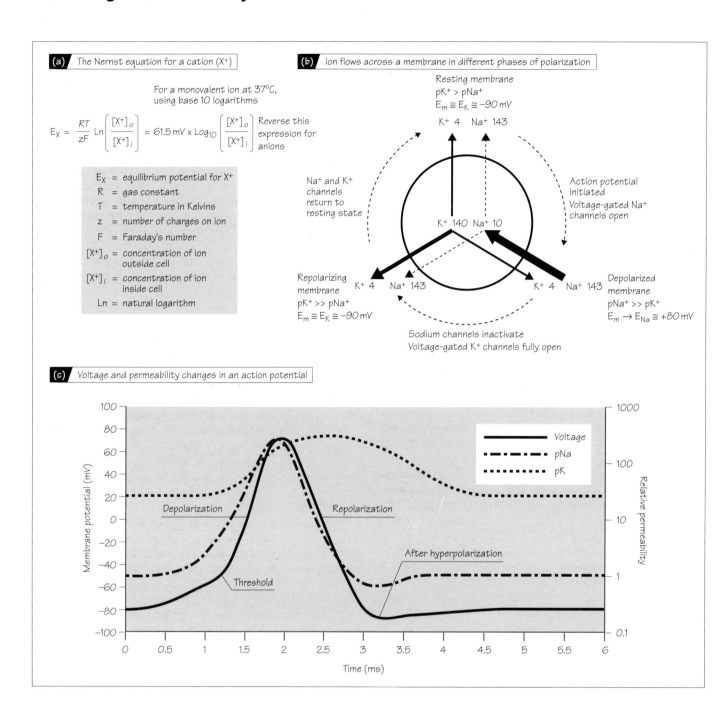

(a) The Nernst equation for a cation (X⁺)

For a monovalent ion at 37°C, using base 10 logarithms

$$E_X = \frac{RT}{zF} \ln\left(\frac{[X^+]_o}{[X^+]_i}\right) = 61.5\,mV \times \log_{10}\left(\frac{[X^+]_o}{[X^+]_i}\right)$$ Reverse this expression for anions

E_X = equilibrium potential for X⁺
R = gas constant
T = temperature in Kelvins
z = number of charges on ion
F = Faraday's number
$[X^+]_o$ = concentration of ion outside cell
$[X^+]_i$ = concentration of ion inside cell
Ln = natural logarithm

(b) Ion flows across a membrane in different phases of polarization

Resting membrane
pK⁺ > pNa⁺
$E_m \cong E_K \cong -90\,mV$

K⁺ 4   Na⁺ 143

Na⁺ and K⁺ channels return to resting state

Action potential initiated
Voltage-gated Na⁺ channels open

K⁺ 140 Na⁺ 10

Repolarizing membrane
pK⁺ >> pNa⁺
$E_m \cong E_K \cong -90\,mV$

K⁺ 4   Na⁺ 143

K⁺ 4 Na⁺ 143

Depolarized membrane
pNa⁺ >> pK⁺
$E_m \to E_{Na} \cong +80\,mV$

Sodium channels inactivate
Voltage-gated K⁺ channels fully open

(c) Voltage and permeability changes in an action potential

Depolarization

Repolarization

Threshold

After hyperpolarization

Voltage
pNa
pK

Membrane potential (mV)

Relative permeability

Time (ms)

Electrical events in biological tissues are caused by the movement of ions across the plasma membrane. A potential difference exists across the membranes of all cells (the **membrane potential, E_m**), but only specialized tissues (known collectively as **excitable tissues**) can generate meaningful electrical signals (**action potentials**). Action potentials transmit information in nerve cells (Chapter 7) and trigger contractions in muscle cells (Chapter 50). Cell membranes are electrically polarized so that the inside is negative relative to the outside. In non-excitable tissues, E_m is roughly $-10\,mV$, whereas in excitable cells it is usually between -60 and $-90\,mV$. The large E_m value of nerve and muscle cells is a prerequisite for the generation of action potentials.

The resting membrane potential

E_m is generated from the asymmetrical distribution of ions, particularly Na^+ and K^+, across the membrane and the differential permeability of the membrane to these ions (Chapters 3 & 5). The concentrations of Na^+ and K^+ ions inside and outside the cell (Chapter 3) are essentially fixed and are altered only in periods of very intense electrical activity. In resting cell membranes (i.e. those not undergoing an action potential), there are more K^+ than Na^+ channels in an open state, so that the permeability to K^+ ions exceeds that for Na^+. In non-excitable cells, the $K^+:Na^+$ permeability ratio is $2:1$, but in nerve and muscle cells it is closer to $25:1$. Ions move through open channels under the influence of chemical concentration gradients and electrical potential differences. These forces can reach an equilibrium, and the potential difference necessary to balance a given ionic concentration gradient can be calculated from the **Nernst equation** (Fig. 6a). Substituting the known concentrations of K^+ and Na^+ into this equation tells us that the voltage required to balance the concentration gradient for K^+ (the **equilibrium potential** for K^+, E_K) is close to $-90\,mV$, whereas that needed to counteract the gradient acting on Na^+ (E_{Na}) is $+80\,mV$. Thus, the E_m value depends primarily on the ratio of the permeabilities for K^+ and Na^+ ions. If all K^+ channels are open and all Na^+ channels are closed, which is close to the situation in excitable cells, E_m equals E_K (Fig. 6b). If the situation is reversed, then E_m equals E_{Na}. Values between these extremes, as found in non-excitable cells, are obtained by opening differing relative numbers of K^+ and Na^+ channels.

The action potential

In excitable cells at rest, K^+ ions diffuse across the membrane to the point at which the negative potential attracting them into the cell is balanced by the chemical gradient driving them out. However, the low conductance for Na^+ prevents these ions from reaching electrochemical equilibrium and holds them in an unstable condition in which both electrical and chemical forces are trying to drive them into the cell. In an action potential, E_m becomes positive for about 1 ms (**depolarization**) and is restored to negativity after a further 1–2 ms (**repolarization**; Fig. 6c). Once triggered, an action potential will travel over the entire surface of an excitable cell (it is **propagated**) and will always have the same amplitude (it is **all-or-nothing**). Action potentials are most often initiated by the opening of ligand-gated ion channels (Chapter 5) that increase the permeability of the membrane to Na^+, allowing these ions to enter the cell and thus cause E_m to move towards E_{Na} (i.e. become positive; Fig. 6b,c). Such a shift in potential opens voltage-gated sodium channels (Chapter 5), further increasing the permeability to that ion and inducing even more depolarization. When E_m reaches -50 to $-45\,mV$ (**threshold potential**), this process becomes explosively self-regenerating and leads to a large, transient increase in permeability to Na^+, giving an E_m value that approaches E_{Na}. This is the positive 'spike' of the action potential (Fig. 6b,c). At the peak of depolarization, Na^+ channels inactivate and the permeability to Na^+ plummets. At the same time, slowly activating, voltage-gated K^+ channels open, so that E_m moves back towards E_K. The Na^+ channels are inactive for about 1 ms and, during this period, they cannot be opened by any amount of depolarization. This is known as the **absolute refractory period** during which it is impossible to generate another action potential. For the following 2–3 ms, the high permeability of the membrane to K^+ ions renders it more difficult to depolarize, an interval known as the **relative refractory period**, when an action potential can be generated only in response to a larger than normal stimulus. The refractory period limits the frequency at which action potentials can be generated to $<1000\,s^{-1}$ and ensures that, once initiated, an action potential can travel only in one direction. The very small changes in ion concentrations that occur during an action potential are restored by the action of the sodium–potassium ATPase.

7 Conduction of action potentials

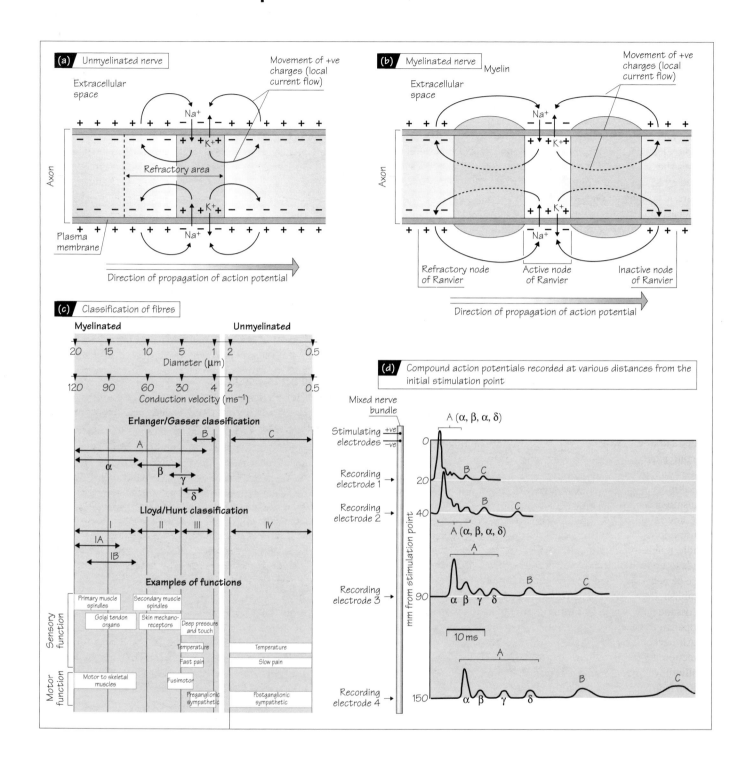

(a) Unmyelinated nerve

Movement of +ve charges (local current flow)

Extracellular space

Na⁺

K⁺

Refractory area

Axon

K⁺

Plasma membrane

Na⁺

Direction of propagation of action potential

(b) Myelinated nerve — Myelin

Movement of +ve charges (local current flow)

Extracellular space

Na⁺

K⁺

Axon

K⁺

Na⁺

Refractory node of Ranvier — Active node of Ranvier — Inactive node of Ranvier

Direction of propagation of action potential

(c) Classification of fibres

Myelinated | Unmyelinated

Diameter (μm): 20 15 10 5 1 2 0.5

Conduction velocity (ms⁻¹): 120 90 60 30 4 2 0.5

Erlanger/Gasser classification

A — B — C

α β γ δ

Lloyd/Hunt classification

I — II — III — IV

IA IB

Examples of functions

Sensory function:
- Primary muscle spindles
- Secondary muscle spindles
- Golgi tendon organs
- Skin mechano-receptors
- Deep pressure and touch
- Temperature
- Temperature
- Fast pain
- Slow pain

Motor function:
- Motor to skeletal muscles
- Fusimotor
- Preganglionic sympathetic
- Postganglionic sympathetic

(d) Compound action potentials recorded at various distances from the initial stimulation point

Mixed nerve bundle

Stimulating electrodes (+ve, –ve)

A (α, β, α, δ)

Recording electrode 1 — 0, 20

B C

Recording electrode 2 — 40

B C

A (α, β, α, δ)

Recording electrode 3 — 90

A

α β γ δ

B C

10 ms

Recording electrode 4 — 150

A

α β γ δ

B C

mm from stimulation point

22 **Introduction**

The **action potential** described in Chapter 6 is a local event that can occur in all excitable cells. This local event is an **all-or-nothing** response, leading to a change in polarity from negative on the inside of the cell (−70 mV) with respect to the outside. This polarity is abolished and reversed (+40 mV) for a short time during the course of the action potential.

Local currents are set up around the action potential because the positive charges from the membrane ahead of the action potential are drawn towards the area of negativity surrounding the action potential (current sink). This decreases the polarity of the membrane ahead of the action potential.

This electronic depolarization initiates a local response that causes the opening of the voltage-gated ion channels (Na^+ followed by K^+); when the threshold for firing of the action potential occurs, it **propagates** the action potential which, in turn, leads to the local depolarization of the next area, and so on. Once initiated, an action potential does not depolarize the area behind it sufficiently to initiate another action potential because the area is **refractory** (see Chapter 6).

This successive depolarization moves along each segment of an **unmyelinated** nerve until it reaches the end. It is all-or-nothing and does not decrease in size (Fig. 7a).

Saltatory conduction

Conduction in **myelinated** axons depends on a similar pattern of current flows. However, because **myelin** is an insulator and because the membrane below it cannot be depolarized, the only areas of the myelinated axon that can be depolarized are those that are devoid of any myelin, i.e. at the **nodes of Ranvier.**

The depolarization jumps from one node to another and is called **saltatory**, from the Latin to jump (*saltare*) (Fig. 7b). Saltatory conduction is rapid and can be up to 50 times faster than in the fastest unmyelinated fibres.

Saltatory conduction not only increases the velocity of impulse transmission by causing the depolarization process to jump from one node to the next, but also **conserves energy** for the axon because depolarization only occurs at the nodes and not along the whole length of the nerve fibre, as in unmyelinated fibres. This leads to up to 100 times less movement of ions than would otherwise be necessary, therefore conserving the energy required to re-establish the sodium and potassium concentration differences across the membranes following a series of action potentials being propagated along the fibre.

All nerve fibres are capable of conducting impulses in either direction if stimulated in the middle of their axon; however, normally they conduct impulses in one direction only (**orthodromically**), from either the receptor to the axon terminal or from the synaptic junction to the axon terminal. **Antidromic** conduction does not normally occur.

Fibre diameters and conduction velocities

Some information needs to be transmitted to and from the central nervous system very rapidly, whilst other information does not. Nerve fibres are able to cover both of these extremes and any in between by virtue of their size, and therefore conduction velocity, and whether or not they are myelinated. Nerve fibres come in all sizes, from 0.5 to 20 μm in diameter, with the smallest diameter unmyelinated fibres being the slowest conducting and the largest myelinated fibres the fastest conducting.

Classification of nerve fibres

Unfortunately, there are two classifications of nerve fibres. One, originally described by Erlanger and Gasser, and often referred to as the **General** classification, uses the letters **A, B and C, with A further subdivided into α, β, γ and δ.** The second, originally described by Lloyd and Hunt, and often referred to as the **Sensory** or **Afferent** classification, uses the Roman numerals **I, II, III and IV, with I further subdivided into a and b.** The groups are subdivided differently in the two classifications and so, unfortunately, it is not possible to rely on only one of the classifications for the description of nerve fibres. **The fibres of groups A and B and also of groups I, II and III are all myelinated and those of group C and IV are unmyelinated.** A table of these classifications, conduction velocities, fibre diameters and examples of their functions is shown in Fig. 7c.

Compound action potentials

Peripheral nerves in most animals comprise a number of axons bound together by a fibrous tissue called the **epineurium.** When extracellular recording electrodes are placed close to a peripheral nerve, the recorded voltage signal, when an action potential is initiated in the bundle, is much smaller (microvolts) than that recorded by an electrode inserted directly into the axon (millivolts). The extracellular recorded signal is made up of the electrical events occurring in all of the active fibres within the nerve bundle. If all the nerve fibres in a nerve bundle are synchronously stimulated at one end of a nerve, and recording electrodes are placed at a number of locations along the length of the bundle, a **compound action potential** is recorded at each electrode. The waveform recorded from each of the electrodes will differ due to the different conduction velocities of each group of fibres that make up the bundle.

Theoretically, if the nerve bundle were to contain examples of all classifications of nerve fibres (i.e. Aα, Aβ, Aγ, Aδ, B and C fibres), the recorded compound action potential would be seen as a **multipeaked** display, as the action potentials in the fastest conducting fibres (Aα) would reach the electrode before those in the slowest conducting fibres (C). Action potentials in the fibres with conduction velocities between these two extremes would arrive between these two times (Fig. 7d).

8 The autonomic nervous system

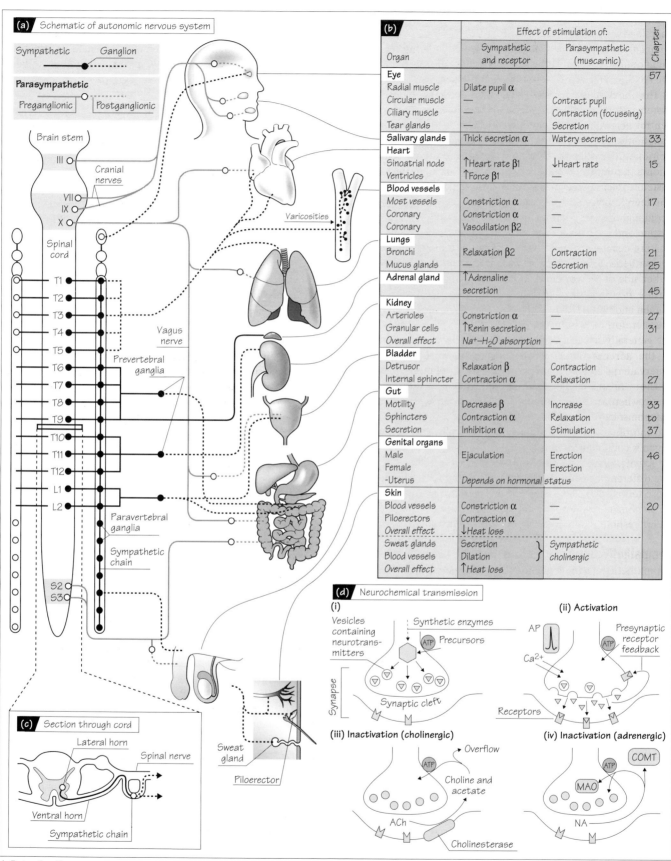

(a) Schematic of autonomic nervous system

Sympathetic — Ganglion
Parasympathetic — Preganglionic — Postganglionic
Brain stem
III, VII, IX, X — Cranial nerves
Spinal cord
T1–T12, L1, L2
S2, S3
Prevertebral ganglia
Paravertebral ganglia
Sympathetic chain
Vagus nerve
Varicosities

(b)

Organ	Effect of stimulation of: Sympathetic and receptor	Parasympathetic (muscarinic)	Chapter
Eye			57
Radial muscle	Dilate pupil α		
Circular muscle	—	Contract pupil	
Ciliary muscle	—	Contraction (focussing)	
Tear glands	—	Secretion	
Salivary glands	Thick secretion α	Watery secretion	33
Heart			15
Sinoatrial node	↑Heart rate β1	↓Heart rate	
Ventricles	↑Force β1	—	
Blood vessels			17
Most vessels	Constriction α	—	
Coronary	Constriction α	—	
Coronary	Vasodilation β2	—	
Lungs			
Bronchi	Relaxation β2	Contraction	21
Mucus glands	—	Secretion	25
Adrenal gland	↑Adrenaline secretion		45
Kidney			
Arterioles	Constriction α	—	27
Granular cells	↑Renin secretion		31
Overall effect	Na⁺–H₂O absorption	—	
Bladder			
Detrusor	Relaxation β	Contraction	
Internal sphincter	Contraction α	Relaxation	27
Gut			33
Motility	Decrease β	Increase	to
Sphincters	Contraction α	Relaxation	37
Secretion	Inhibition α	Stimulation	
Genital organs			46
Male	Ejaculation	Erection	
Female		Erection	
-Uterus	Depends on hormonal status		
Skin			20
Blood vessels	Constriction α	—	
Piloerectors	Contraction α	—	
Overall effect	↓Heat loss		
Sweat glands	Secretion	} Sympathetic cholinergic	
Blood vessels	Dilation		
Overall effect	↑Heat loss		

(Na⁺–H₂O = Na^+–H_2O)

(c) Section through cord

Lateral horn
Spinal nerve
Spinal cord
Ventral horn
Sympathetic chain
Sweat gland
Piloerector

(d) Neurochemical transmission

(i)
Vesicles containing neurotransmitters
Synthetic enzymes
ATP Precursors
Synapse
Synaptic cleft

(ii) Activation
AP
Ca²⁺
Presynaptic receptor feedback
ATP
Receptors

(iii) Inactivation (cholinergic)
ATP
Overflow
Choline and acetate
ACh
Cholinesterase

(iv) Inactivation (adrenergic)
ATP
COMT
MAO
NA

The **autonomic nervous system (ANS)** provides the **efferent** pathway for the **involuntary** control of most organs, excluding the motor control of skeletal muscle (Chapters 50 & 51). The ANS provides the effector arm for **homeostatic reflexes** (e.g. control of blood pressure), and allows the integration and modulation of function by central mechanisms in the brain in response to *environmental* and *emotional* stimuli (e.g. exercise, thermoregulation, 'fight or flight'). Figure 8a shows a simplified schematic diagram of the ANS, and Fig. 8b its actions on major organs.

The ANS is divided into **sympathetic** and **parasympathetic** systems. Both contain **preganglionic neurones** originating in the central nervous system that synapse with non-myelinated **postganglionic neurones** in the **peripheral ganglia**; postganglionic neurones innervate the target organ or tissue (Fig. 8a,b). Preganglionic neurones of both sympathetic and parasympathetic systems release **acetylcholine** in the synapse, which acts on **cholinergic nicotinic** receptors on the postganglionic fibre. The postganglionic neurotransmitters and receptors depend on the system and organ (see below). Parasympathetic peripheral ganglia are generally found close to or in the target organ, whereas sympathetic ganglia are largely located in two **sympathetic chains** either side of the vertebral column (*paravertebral ganglia*), or in diffuse *prevertebral ganglia* of the visceral plexuses of the abdomen and pelvis (Fig. 8a). Sympathetic postganglionic neurones are therefore generally long, whereas parasympathetic neurones are generally short. An exception is the sympathetic innervation of the **adrenal gland**, where preganglionic neurones directly innervate the adrenal medulla.

The sympathetic system is more pervasive than the parasympathetic; where an organ is innervated by both systems, they often act antagonistically (Fig. 8b). However, there is a high degree of central coordination, so that an increase in sympathetic activity to an organ is commonly accompanied by a decrease in parasympathetic activity. Sympathetic and parasympathetic activity may modulate different functions in the same organ (e.g. genital organs). In loose terms, the sympathetic system might be said to coordinate '*flight or fight*' responses, and the parasympathetic system '*rest and digest*' responses.

Sympathetic system

Sympathetic preganglionic neurones originate in the *lateral horn* of segments T1–L2 of the spinal cord, and exit the cord via the *ventral horn* (Fig. 8c) on their way to the paravertebral or prevertebral ganglia. Sympathetic postganglionic neurones terminate in the effector organs, where they release **noradrenaline (norepinephrine)**. Noradrenaline and **adrenaline (epinephrine)**, which is released by the adrenal medulla, are catecholamines, and activate **adrenergic** receptors, which are linked via G-proteins to cellular effector mechanisms. There are two main classes of adrenergic receptor, α and β, and these are further subdivided into several subtypes (e.g. α_1, α_2, β_1, β_2). Noradrenaline and adrenaline are equally potent on α_1-receptors, which are linked to G_q-proteins and are commonly associated with smooth muscle contraction (e.g. blood vessels). The α_2-receptors are $G_{i/o}$-protein linked and are often inhibitory. All β-receptors are linked to G_s-protein and activate adenylyl cyclase to make cyclic adenosine monophosphate (cAMP). Noradrenaline is more potent at β_1-receptors and adrenaline is more potent at β_2-receptors. The activation of β-receptors is associated with the relaxation of smooth muscle (e.g. blood vessels, airways), but causes increased heart rate and force (Fig. 8b).

A few sympathetic neurones release acetylcholine at the effector (e.g. sweat glands), and are thus known as **sympathetic cholinergic** neurones.

Parasympathetic system

Parasympathetic preganglionic neurones originate in the brain stem, from which they run in the IIIrd, VIIth, IXth and Xth (**vagus**) cranial nerves, and also from the second and third sacral segments of the spinal cord (Fig. 8a). Parasympathetic postganglionic neurones release acetylcholine, which acts on **cholinergic muscarinic** receptors. Parasympathetic activation causes secretion in many glands (e.g. bronchial mucous glands), and either contraction (e.g. bladder detrusor) or relaxation (e.g. bladder internal sphincter) of smooth muscle, although it has little effect on blood vessels. Notable exceptions, however, include vasodilatation in the penis and clitoris with subsequent erection (Chapter 45).

Neurochemical transmission

Action potentials (APs) in incoming neurones are transmitted by the release of neurotransmitters that bind to receptors on the postganglionic neurone or effector tissue. Between neurones (e.g. in ganglia), this occurs within a classical **synapse**, where the axon terminates in a bulbous swelling or **bouton** separated from the target by a narrow (10–20 nm) synaptic cleft (Fig. 8di). Postganglionic neurones branch repeatedly and have numerous boutons along their length, forming **varicosities** (e.g. see blood vessel in Fig. 8a). The boutons may either be close (~20 nm) to the effector membrane, allowing fast and specific delivery of the signal, or at some distance (100–200 nm), allowing a more distributed but slower effect. The mechanisms of neurochemical transmission are similar, and although the text below and Fig. 8di–iv refer to synapses, the same principles apply.

Synthetic enzymes are transported down the axon into the bouton, where they synthesize neurotransmitter (acetylcholine, noradrenaline) from precursors transported into the bouton. The neurotransmitter is stored in 50 nm **vesicles** (Fig. 8di). The arrival of an AP at the nerve ending causes an influx of Ca^{2+}, the fusion of vesicles with the membrane and the release of neurotransmitter; this binds to postsynaptic receptors and activates the response. Neurotransmitter release can be suppressed by feedback onto **presynaptic inhibitory receptors** (α_2-receptors for adrenergic synapses) (Fig. 8dii). Neurotransmitters must be removed at the end of activation. In *cholinergic* synapses, **cholinesterase** rapidly breaks down acetylcholine into *choline* and *acetate*, which are recycled; some may escape into interstitial fluid (*overflow*) (Fig. 8diii). In adrenergic synapses, most noradrenaline is rapidly taken up again by the nerve ending via an adenosine triphosphate (ATP)-dependent transporter called **uptake-1**; recovered noradrenaline is recycled. Some facilitated diffusion (**uptake-2**) also occurs into smooth muscle. Excess noradrenaline and sympathomimetic amines, such as tyramine (found in some foodstuffs), are metabolized in the neurone by mitochondrial **mono-amine oxidase (MAO)**. Noradrenaline and other catecholamines that enter the circulation are metabolized sequentially by **catechol-O-methyl transferase (COMT)** and MAO (Fig. 8div).

9 Blood

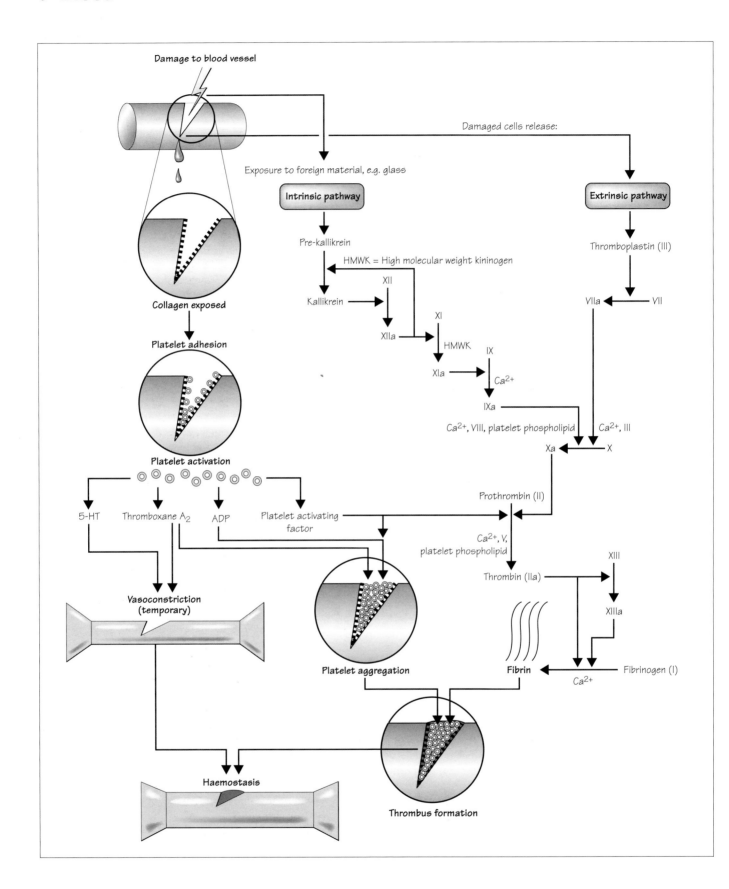

Diffusion is the primary method of moving nutrients, waste products and chemical signals between cells. As diffusion is rapid only over small distances (Chapter 5), multicellular organisms, such as ourselves, require a transport system to carry materials to and from the cells. The evolutionary answer to this problem has been the development of a cardiovascular system (Chapter 12) that carries a transport medium, **blood**, to within a few micrometres of all cells. The main constituents of blood are the liquid component, **plasma** (Chapter 3), and red and white blood cells. **Red blood cells (erythrocytes)** are specialized to transport respiratory gases and contain haemoglobin (Chapter 24), whereas **white cells (leukocytes)** are the active part of the body's defence systems (Chapter 10). Normal values for red and white cell counts, haemoglobin, and the proportion of blood volume due to red cells (**haematocrit** or **packed cell volume**; estimated by spinning down a blood sample) are shown in Table 9.

Table 9 Average adult values for whole blood

	Males	Females
Red cell count (L^{-1})	4.5–5.5×10^{12}	3.8–4.8×10^{12}
Haemoglobin concentration ($g\,L^{-1}$)	130–170	120–150
Haematocrit	0.40–0.50	0.36–0.46
Platelet count (L^{-1})	1.4–4.0×10^{11}	
White cell counts (L^{-1})	4–11×10^{9}	

Red blood cells

Red cells are roughly 8 µm across at their widest point, have a biconcave shape and, uniquely, possess no nucleus. They produce one protein, the oxygen-carrying molecule **haemoglobin** (Chapter 20), made from messenger ribonucleic acid (mRNA) templates that remain after the nucleus is removed during development of the cell. The lack of a nucleus means that they cannot repair themselves, and red cells have a lifespan of just 100–130 days, after which time they are removed from the circulation by the liver and the spleen. In adults, red cells are made in the **red marrow** of the main trunk bones, the skull and the ends of the humerus and femur. In juveniles, all large bones make red blood cells, whereas, in neonates, the process occurs in the liver and the spleen. Red cell production is stimulated by **erythropoietin (EPO)**, a hormone released from the kidney in response to low blood oxygen levels (**hypoxia**). Red cells in excess of immediate requirements are stored in the **spleen** and are released during episodes of hypoxia or haemorrhage. The production of haemoglobin is dependent on supplies of **iron**, **folate** and **vitamin B_{12}** all of which are obtained from the diet. An inadequate supply of these factors, either through a lack of intake or poor absorption (Chapter 35), compromises the oxygen-carrying capacity of the blood and leads to the condition known as **anaemia** (Chapter 24). Anaemia can also arise from a reduced red cell count (e.g. from haemorrhage or white cell cancers, e.g. **leukaemia**) or abnormalities of haemoglobin as in **thalassaemia**, or the **sickle cell** mutation that protects against malaria.

White blood cells and plasma

White blood cells include **lymphocytes** (a key part of the immune system; Chapter 10), **monocytes** (which on entering the tissues become **macrophages**), and granulocytes. Granulocytes (**neutrophils, eosinophils** and **basophils**) destroy bacteria by **phagocytosis** (engulfing them) and also release mediators such as histamine that are important in inflammation. The major biological properties of plasma derive from its protein content. The main plasma protein is **albumin**, which provides most of the plasma **oncotic pressure** (Chapters 3 & 19) and binds several hormones, bile pigments (Chapter 36) and Ca^{2+} ions for transport in the blood. Other important proteins are the globulins (involved in the immune response; Chapter 10) and **fibrinogen**, an element of the haemostatic cascade (Fig. 9). Plasma proteins are produced in the liver.

Platelets and haemostasis

Blood is a precious resource, and so leaks in the circulation must be sealed quickly to stop bleeding by the process of **haemostasis**. This is achieved by a series of mechanisms that culminate in the production of a **blood clot**. Platelets, small (2–4 µm) vesicle-like fragments produced from large cells known as **megakaryocytes**, are very important in haemostasis. The initial response to blood vessel damage involves adhesion of **platelets** to exposed collagen, a structural protein in the vessel wall. This activates the platelets, which release **5-hydroxytryptamine (5-HT**, or **serotonin**) and **thromboxane A_2** (an eicosanoid; see chapter 10) causing **vasoconstriction**, whereby the smooth muscle of the vessel wall contracts to limit blood flow, and also allows the clot to develop without being washed away. Activated platelets also release adenosine diphosphate (ADP) and platelet activating factor (PAF), which together with thromboxane attract more platelets to the injury site (**aggregation**) where they form a soft plug. This plug is then strengthened with **fibrin**, an insoluble filamentous protein that binds the platelets together. Fibrin is made from plasma fibrinogen by the enzyme **thrombin**. Activation of thrombin and production of fibrin are the final steps in the haemostatic cascade (Fig. 9), a process involving 12 different proteins (**clotting factors**). Most of these proteins are designated by roman numerals and all are manufactured in the liver, which in some cases (factors II, VII, IX and X) requires vitamin K. Clotting factors exist in blood as inactive pro-enzymes that are activated during coagulation (Fig. 9). Clot formation involves an 'intrinsic' pathway, which is activated by collagen and 'foreign' materials such as glass or metal, and an 'extrinsic' pathway, stimulated by **thromboplastin** (factor III), a protein released from damaged cells. The pathways converge with the activation of factor X, which, in turn, stimulates the conversion of prothrombin to thrombin and thus production of fibrin (Fig. 9). PAF released from platelets also activates thrombin. A clot is stable for several days and is eventually removed by the enzyme **plasmin** after the tissue has been repaired.

Haemophilia is a sex-linked (i.e. only males are susceptible) genetic disorder in which one of the factors VIII or IX is low or absent. Haemophiliacs show the normal vasoconstrictor response to tissue damage, but cannot produce a stable blood clot to stop bleeding. The condition is currently treated using replacement factors produced by genetically engineered bacteria.

10 Inflammation and immunity

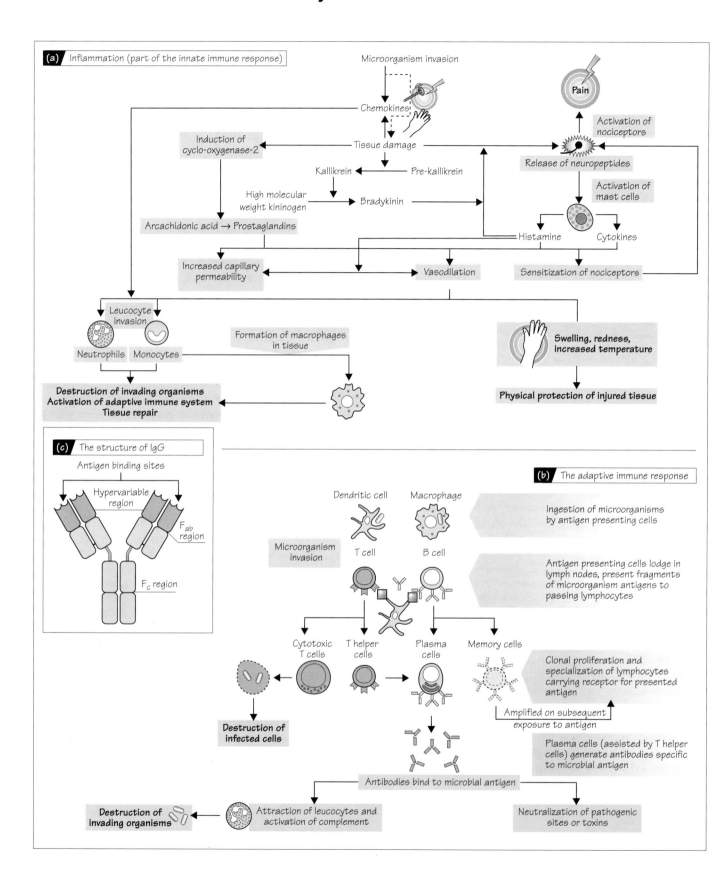

The haemostatic system prevents blood loss and, by the formation of clots and scabs, provides a protective covering where the skin has been injured. Damage to the external surfaces of the body provides routes of entry for foreign materials, including pathogenic microorganisms. The immune system combats invasion by pathogens in two ways: an innate, rapidly activated and non-specific response that often manifests as **inflammation**, and **adaptive immunity**, whereby immune cells produce antibodies in response to the presence of foreign proteins.

Inflammation: the innate immune response

Tissue damage activates a number of proteins, one of which is the enzyme **kallikrein** that catalyses the cleavage of **bradykinin** from high-molecular-weight kininogen (HMWK; Chapter 9). In addition, the production of the tissue enzyme **cyclo-oxygenase-2** is induced at the site of injury to catalyse the conversion of arachidonic acid (a product of membrane phospholipids; Chapter 4) into a series of local signalling molecules known as **eicosanoids**. Of these, the most important inflammatory mediators are the **prostaglandins**. Mast cells and macrophages present in the affected tissue release **histamine** and **cytokines**, of which the peptides **interleukin-1β** (**IL-1β**) and **tumour necrosis factor** (**TNF**) are crucial. Bradykinin, prostaglandins, cytokines and histamine are prime mediators in acute inflammation, although many other factors are also involved. Collectively, these mediators cause: (i) dilation of the blood vessels at the site of injury to cause reddening and an increase in local temperature; (ii) the capillary walls to become permeable to plasma proteins to cause swelling; and (iii) the sensitization or direct activation of the terminals of nociceptors (Chapter 55) to cause pain. Pain sensation encourages guarding behaviour and swelling provides mechanical protection to the damaged tissue. Leucocytes, white blood cells that include **neutrophils** and **monocytes**, are attracted to the site of injury by **chemokines** (signal molecules) released from inflamed tissue. Leucocytes engulf and kill (**phagocytose**) invading organisms. Monocytes can leave the circulation to become tissue **macrophages**, large cells that kill microorganisms and remove damaged tissue as the wound heals (Fig. 10a). **Natural killer cells** are white cells that are part of the innate immune system, but do not normally contribute to inflammation. They are able to target and destroy any abnormal (cancerous or virally infected) cells that they encounter.

Adaptive immunity

Innate responses provide a non-specific defence of the body against all invading organisms. In contrast, the adaptive immune response is tailored to destroy particular organisms by recognition of marker molecules (usually proteins), known as **antigens**, that can be targeted by specific **antibodies** (Fig. 10b,c). The system normally differentiates between host molecules (self) and foreign antigens (non-self). The effectors of immunity are the white blood cells known as **lymphocytes**, which exist in two main forms: T and

B cells. Both types derive from bone marrow, with T cells, which account for some 80% of lymphocytes, undergoing important developmental stages in the **thymus**. Lymphocytes continuously recirculate between the blood and the **lymphoid tissues** (the lymph nodes, specialized tissues in the gut and the spleen) to ensure that a lymphocyte will find foreign antigens quickly no matter where they enter the body. All lymphocytes carry cell surface receptors that recognize antigens, but each individual cell will respond *to a single antigen only*. There are an estimated 10^8–10^9 different lymphocyte antigen receptors that have evolved to account for virtually all of the foreign antigens that we are likely to meet. The system responds to novel antigens when they are delivered to it by **antigen-presenting cells**, the most important of which are the **dendritic cells** and **macrophages**. These white blood cells collect foreign antigens from sites of infection and carry them to the lymphoid tissues, where they are exposed to the flow of circulating lymphocytes. T and B cells carrying receptors for the presented antigen (and *only* those cells) undergo clonal proliferation. Some of the activated T cells (**cytotoxic T cells**) will kill any infected cells that they encounter. Other T cells (**T helper cells**) facilitate the production, by activated B cells, of antibodies specific to the antigen (see below; Fig. 10b). This system takes some 5 days to respond fully to initial exposure to an antigen, but some of the stimulated lymphocytes differentiate into **memory cells** that enable a much faster response to be made to the antigen should the pathogen invade the body on subsequent occasions. This is the basis of **active immunization**. (NB. This is in contrast to **passive** immunization, in which premanufactured antibodies are delivered directly into the patient.) HIV (**human immunodeficiency virus**), the pathogenic organism in AIDS (**acquired immune deficiency syndrome**), selectively kills T helper cells and massively reduces the effectiveness of the immune system.

Antibodies

B cells stimulated by antigen develop into **plasma cells** that lodge in the lymph nodes. These cells produce antibodies that specifically bind the stimulating antigen, having the effects of: (i) attracting and facilitating the action of phagocytic leucocytes; (ii) neutralizing toxins; (iii) blocking pathogenic sites on virus proteins; and (iv) activating **complement**, an enzyme cascade that has pro-inflammatory actions and which can itself lyse some microorganisms. Antibodies exist in several forms, the most common (80%) of which is immunoglobulin G (IgG) (Fig. 10c), a γ-globulin comprising a constant (F_c) stem with hypervariable (F_{ab}) branches that provide specificity. Immunoglobulin M (IgM) is produced by immature B cells when they first respond to an antigen and is a potent stimulator of complement. Immunoglobulin A (IgA) antibodies are specialized for release in exocrine secretions, such as saliva, tears and milk, where they bind extracellular pathogens and toxins. Finally, immunoglobulin E (IgE) attaches to mast cells so that they can be activated in response to specific antigens, a reaction that is involved in **allergic responses**.

11 Principles of diffusion and flow

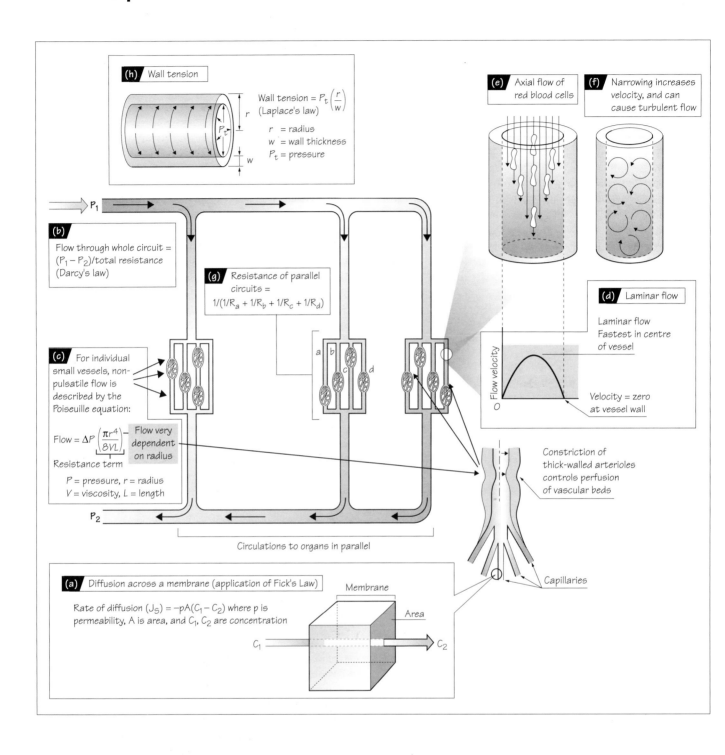

(h) Wall tension

$$\text{Wall tension} = P_t\left(\frac{r}{w}\right)$$
(Laplace's law)

r = radius
w = wall thickness
P_t = pressure

(e) Axial flow of red blood cells

(f) Narrowing increases velocity, and can cause turbulent flow

P_1

(b) Flow through whole circuit = $(P_1 - P_2)$/total resistance (Darcy's law)

(g) Resistance of parallel circuits = $1/(1/R_a + 1/R_b + 1/R_c + 1/R_d)$

(d) Laminar flow

Laminar flow Fastest in centre of vessel

Velocity = zero at vessel wall

Flow velocity

O

(c) For individual small vessels, non-pulsatile flow is described by the Poiseuille equation:

$$\text{Flow} = \Delta P\left(\frac{\pi r^4}{8VL}\right)$$

Flow very dependent on radius

Resistance term

P = pressure, r = radius
V = viscosity, L = length

P_2

Circulations to organs in parallel

Constriction of thick-walled arterioles controls perfusion of vascular beds

Capillaries

(a) Diffusion across a membrane (application of Fick's Law)

Rate of diffusion (J_s) = $-pA(C_1 - C_2)$ where p is permeability, A is area, and C_1, C_2 are concentration

Membrane

Area

C_1

C_2

Diffusion and bulk flow

Materials are carried around the body by a combination of bulk flow and diffusion. **Bulk flow** simply means transport *with* the carrying medium (blood, air). **Passive diffusion** refers to movement down a *concentration gradient*, and accounts for transport across small distances, e.g. within the cytosol and across membranes. The rate of **diffusion in a solution** is described by **Fick's law**:

$$J_s = -DA(\Delta C / \Delta x) \tag{11.1}$$

where J_s is the amount of substance transferred per unit time, ΔC is the difference in concentration, Δx is the diffusion distance and A is the surface area over which diffusion occurs. The negative sign reflects movement *down* the concentration gradient. D is the **diffusion coefficient**, a measure of how easy it is for the substance to diffuse. D is related to temperature, solvent viscosity and the size of the molecule, and is normally *inversely proportional to the cube root of the molecular weight*. **Diffusion across a membrane** is affected by the **permeability** of the membrane. The permeability (p) is related to the membrane thickness and composition, and the diffusion coefficient of the substance. Fick's equation can be rewritten as:

$$J_s = -pA\Delta C \tag{11.2}$$

where A is the membrane area and ΔC is the concentration difference across the membrane. The rate of diffusion across a capillary wall is therefore related to the *concentration difference* across the wall and the *permeability* of the wall to that substance (Fig. 11a).

Flow through a tube

Flow through a tube is dependent on the pressure difference across the ends of the tube ($P_1 - P_2$) and the resistance to flow provided by the tube (R):

$$Flow = (P_1 - P_2)/R \tag{11.3}$$

This is **Darcy's law** (analogous to *Ohm's law* in electronics) (Fig. 11b).

Resistance is due to frictional forces, and is determined by the **diameter** of the tube and the **viscosity** of the fluid:

$$R = (8VL)/(\pi r^4) \tag{11.4}$$

This is **Poiseuille's law**, where V is the viscosity, L is the length of the tube and r is the radius of the tube. Combining eqns. (11.3) and (11.4) shows an important principle, namely that the **flow ∝ (radius)⁴**:

$$Flow = [(P_1 - P_2)\pi r^4]/(8VL) \tag{11.5}$$

Therefore, small changes in radius have a large effect on flow (Fig. 11c). Thus, the constriction of an artery by 20% will decrease the flow by ~60%.

Viscosity. Treacle flows more slowly than water because it has a higher viscosity. Plasma has a similar viscosity to water, but blood contains cells (mostly erythrocytes) which effectively increase the viscosity by three- to fourfold. Changes in cell number, e.g. *poly-cythaemia* (increased erythrocytes), therefore affect the blood flow.

Laminar and turbulent flow. Frictional forces at the sides of a tube cause drag on the fluid touching them. This creates a *velocity gradient* (Fig. 11d) in which the flow is greatest at the centre. This is termed **laminar flow**, and describes the flow in the majority of cardiovascular and respiratory systems at rest. A consequence of the velocity gradient is that blood cells tend to move away from the sides of the vessel and accumulate towards the centre (**axial streaming**; Fig. 11e); they also tend to align themselves to the flow. In small vessels, this *effectively reduces the blood viscosity* and minimizes the resistance (the **Fåhraeus–Lindqvist effect**).

At high velocities, especially in large arteries and airways, and at edges or branches where the velocity increases sharply, flow may become **turbulent**, and laminar flow is disrupted (Fig. 11f). This significantly increases the resistance. The narrowing of airways and large arteries (or valve orifices), which increases the fluid velocity, can therefore cause turbulence, which is heard as **lung sounds** (e.g. wheezing in asthma) and **cardiac murmurs** (Chapter 14).

Resistances in parallel and series (Fig. 11g). The cardiovascular and respiratory systems contain a mixture of *series* (e.g. arteries ⇒ arterioles ⇒ capillaries ⇒ venules ⇒ veins) and *parallel* (e.g. lots of capillaries) components (Fig. 11g). Flow through a *series* of tubes is restricted by the resistance of each tube in turn, and the total resistance is the **sum** of the resistances:

$$R_T = R_1 + R_2 + R_3 + \ldots \tag{11.6}$$

In a parallel circuit, the addition of extra paths reduces the total resistance, and so:

$$R_T = 1/(1/R_a + 1/R_b \ldots) \tag{11.7}$$

Although the resistance of individual capillaries or terminal bronchioles is high (small radius, *Poiseuille's law*), the huge number of them in parallel means that their contribution to the total resistance of the cardiovascular and respiratory systems is comparatively small.

Wall tension and pressure in spherical or cylindrical containers

Pressure across the wall of a flexible tube (*transmural pressure*) tends to extend it, and increases wall tension. This can be described by **Laplace's law**:

$$P_t = (Tw)/r \tag{11.8}$$

where P_t is the transmural pressure, T is the wall tension, w is the wall thickness and r is the radius (Fig. 11h). Thus, a small bubble with the same wall tension as a larger bubble will contain a greater pressure, and will collapse into the larger bubble if they are joined. In the lung, small alveoli would collapse into larger ones were it not for *surfactant* which reduces the surface tension more strongly as the size of the alveolus decreases (Chapter 22). Laplace's law also means that a large dilated heart (e.g. heart failure) has to develop more wall tension (contractile force) in order to obtain the same ventricular pressure, making it less efficient.

12 Introduction to the cardiovascular system

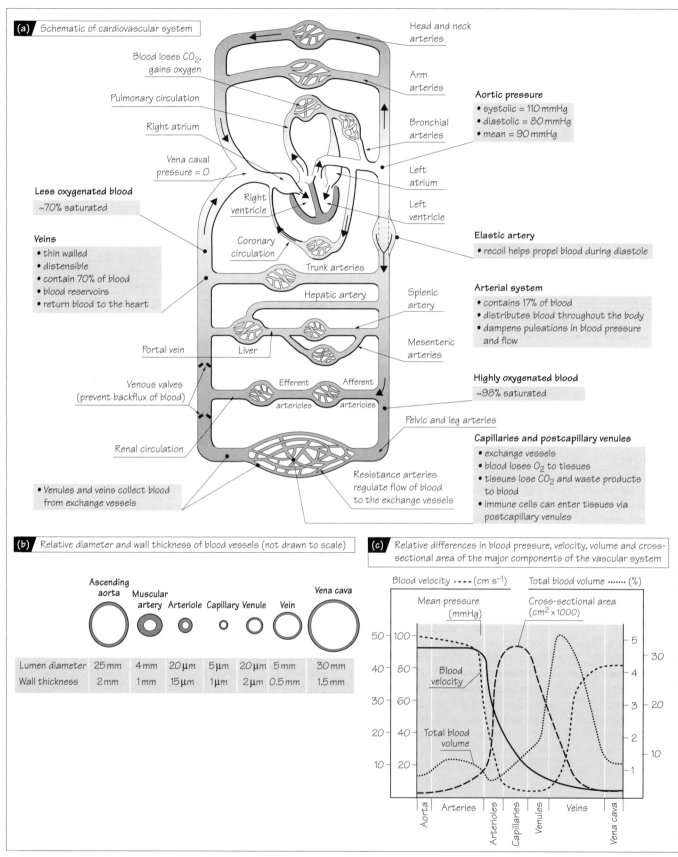

(a) Schematic of cardiovascular system

Head and neck arteries

Blood loses CO_2, gains oxygen

Pulmonary circulation

Right atrium

Arm arteries

Bronchial arteries

Aortic pressure
- systolic = 110 mmHg
- diastolic = 80 mmHg
- mean = 90 mmHg

Vena caval pressure = 0

Left atrium

Left ventricle

Less oxygenated blood
~70% saturated

Right ventricle

Coronary circulation

Trunk arteries

Elastic artery
- recoil helps propel blood during diastole

Veins
- thin walled
- distensible
- contain 70% of blood
- blood reservoirs
- return blood to the heart

Hepatic artery

Splenic artery

Arterial system
- contains 17% of blood
- distributes blood throughout the body
- dampens pulsations in blood pressure and flow

Portal vein

Liver

Mesenteric arteries

Highly oxygenated blood
~98% saturated

Venous valves (prevent backflux of blood)

Efferent arterioles

Afferent arterioles

Pelvic and leg arteries

Renal circulation

Capillaries and postcapillary venules
- exchange vessels
- blood loses O_2 to tissues
- tissues lose CO_2 and waste products to blood
- immune cells can enter tissues via postcapillary venules

- Venules and veins collect blood from exchange vessels

Resistance arteries regulate flow of blood to the exchange vessels

(b) Relative diameter and wall thickness of blood vessels (not drawn to scale)

Ascending aorta
Muscular artery
Arteriole
Capillary
Venule
Vein
Vena cava

	Ascending aorta	Muscular artery	Arteriole	Capillary	Venule	Vein	Vena cava
Lumen diameter	25 mm	4 mm	20 μm	5 μm	20 μm	5 mm	30 mm
Wall thickness	2 mm	1 mm	15 μm	1 μm	2 μm	0.5 mm	1.5 mm

(c) Relative differences in blood pressure, velocity, volume and cross-sectional area of the major components of the vascular system

Blood velocity ---- (cm s^{-1})

Total blood volume ······· (%)

Mean pressure (mmHg)

Cross-sectional area (cm^2 x 1000)

Blood velocity

Total blood volume

Aorta
Arteries
Arterioles
Capillaries
Venules
Veins
Vena cava

The cardiovascular system comprises the heart and blood vessels, and contains ~5.5 L of blood in a 70 kg man. Its main functions are to distribute O_2 and nutrients to tissues, transfer metabolites and CO_2 to excretory organs and the lungs, and transport hormones and components of the immune system. It is also important for thermoregulation. The cardiovascular system is arranged mostly in **parallel**, i.e. each tissue derives blood directly from the aorta (Fig. 12a). This allows all tissues to receive fully oxygenated blood, and flow can be controlled independently in each tissue against a constant pressure head by altering the resistance of small arteries (i.e. arteriolar constriction or dilatation). The right heart, lungs and left heart are arranged in **series**. **Portal systems** are also arranged in series, where blood is used to transport materials directly from one tissue to another, such as the **hepatic portal system** between digestive organs and the liver. The function of the cardiovascular system is modulated by the autonomic nervous system (Chapter 8).

Blood vessels

The vascular system consists of **arteries** that take blood from the heart to the tissues, thin-walled **capillaries** that allow the diffusion of gases and metabolites, and **veins** and **venules** which return blood to the heart. The blood pressure, vessel diameter and wall thickness vary throughout the circulation (Fig. 12b,c). Varying amounts of **smooth muscle** are contained within the vessel walls, allowing them to constrict and alter their resistance to flow (Chapters 11 & 17). Capillaries contain no smooth muscle. The inner surface of all blood vessels is lined with a thin layer of **endothelial cells**, important for vascular function (Chapter 17). Large arteries are **elastic** and partially damp out oscillations in pressure produced by the pumping of the heart; stiff arteries (age, atherosclerosis) result in larger oscillations. Small arteries contain relatively more muscle and are responsible for controlling tissue blood flow. Veins have a larger diameter than equivalent arteries, and provide less resistance. They have thin, distensible walls, and contain ~70% of the total blood volume (Fig. 12c). They are therefore known as **capacitance vessels** and act as a *blood volume reservoir*; when required, they can constrict and effectively increase the blood volume (Chapter 17). Large veins in the limbs contain **one-way valves**, so that when muscle activity (e.g. walking) intermittently compresses these veins they act as a pump, and assist the return of blood to the heart (the **muscle pump**).

The heart

The **heart** is a four-chambered muscular pump which propels blood around the circulation. It has an intrinsic pacemaker and requires no nervous input to beat normally, although it is modulated by the **autonomic nervous system** (Chapter 8). The volume of blood pumped per minute (**cardiac output**) is ~5 L at rest in humans, although this can increase to above 20 L during exercise. The volume ejected per beat (**stroke volume**) is ~70 mL at rest. The **ventricles** perform the work of pumping; **atria** assist ventricular filling. Unidirectional flow through the heart is maintained by **valves** between the chambers and outflow tracts. Contraction of the heart is called **systole**; the period between each systole, when the heart refills with blood, is called **diastole**.

The systemic circulation

During systole, the pressure in the left ventricle increases to ~120 mmHg, and blood is ejected into the **aorta**. The rise in pressure stretches the elastic walls of the aorta and large arteries and drives blood flow. **Systolic pressure** is the maximum arterial pressure during systole (~110 mmHg). During diastole, arterial blood flow is partly maintained by elastic recoil of the walls of large arteries. The minimum pressure reached before the next systole is the **diastolic pressure** (~80 mmHg). The difference between the systolic and diastolic pressures is the **pulse pressure**. Blood pressure is expressed as the systolic/diastolic arterial pressure, e.g. 110/80 mmHg. The **mean blood pressure** (mean arterial pressure, MAP) cannot be calculated by averaging these pressures, because for ~60% of the time the heart is in diastole. It is instead estimated as the *diastolic pressure plus one-third of the pulse pressure*, e.g. $80 + 1/3(110 - 80) \approx 90$ mmHg.

The **major arteries** divide repeatedly into smaller **muscular arteries**, the smallest of which (diameter <100 µm) are called **arterioles**. Tissue blood flow is regulated by the constriction of these small arteries, referred to as **resistance vessels**. The mean blood pressure at the start of the arterioles is ~65 mmHg. The arterioles divide into dense networks of **capillaries** in the tissues, and these rejoin into small and then larger **venules**, the smallest veins. Capillaries and small venules provide the exchange surface between blood and tissues, contain no smooth muscle and are called **exchange vessels**. The pressure on the arterial side of capillaries is ~25 mmHg and, on the venous side, ~15 mmHg. Venules converge into veins and finally the **vena cava**. This returns the partially deoxygenated and CO_2-loaded blood to the right atrium. The pressure in the vena cava at the level of the heart is called the **central venous pressure** (CVP), and is close to 0 mmHg.

The pulmonary circulation

The right atrium helps to fill the right ventricle, which pumps blood into the **pulmonary artery** and lungs. The pulmonary circulation is shorter than the systemic, and has a lower resistance to flow. Less pressure is therefore required to drive blood through the lungs; the pulmonary artery pressure is ~20/15 mmHg. Gas exchange occurs in capillaries surrounding the alveoli (small air sacs) of the lungs. These rejoin to form pulmonary venules and veins, and oxygenated blood is returned through the pulmonary vein to the left atrium, and hence to the left ventricle. The metabolic requirements of the lungs are not met by the pul-monary circulation, but by the separate **bronchial circulation**, the venous outflow of which returns to the left side of the heart (Fig. 12a).

13 The heart

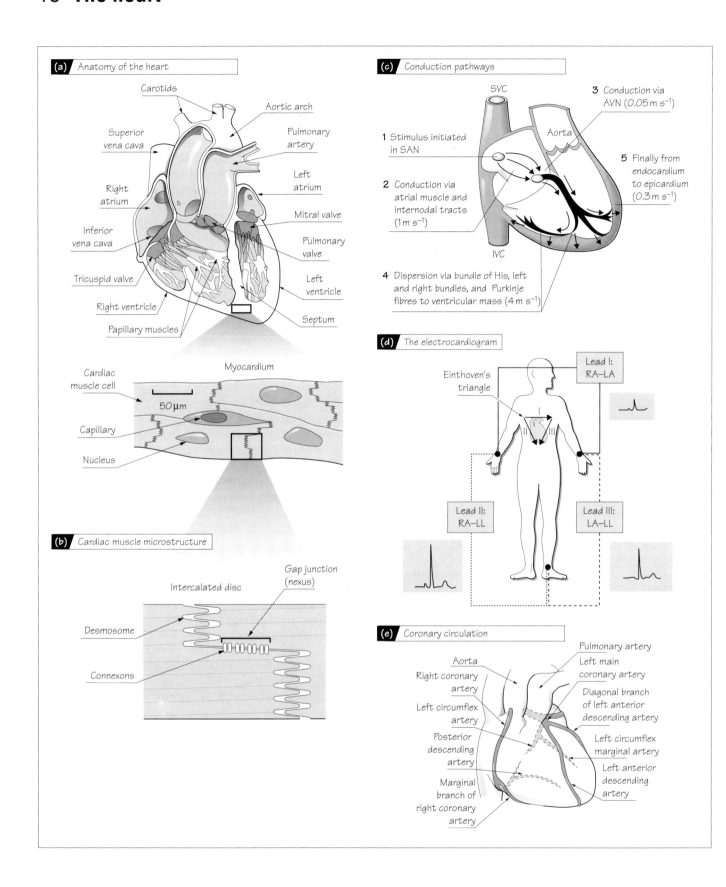

(a) Anatomy of the heart

Carotids
Aortic arch
Superior vena cava
Pulmonary artery
Right atrium
Left atrium
Inferior vena cava
Mitral valve
Tricuspid valve
Pulmonary valve
Right ventricle
Left ventricle
Papillary muscles
Septum

Myocardium
Cardiac muscle cell
50 μm
Capillary
Nucleus

(b) Cardiac muscle microstructure

Intercalated disc
Gap junction (nexus)
Desmosome
Connexons

(c) Conduction pathways

SVC
Aorta
IVC

1 Stimulus initiated in SAN
2 Conduction via atrial muscle and internodal tracts (1m s^{-1})
3 Conduction via AVN (0.05 m s^{-1})
5 Finally from endocardium to epicardium (0.3 m s^{-1})
4 Dispersion via bundle of His, left and right bundles, and Purkinje fibres to ventricular mass (4m s^{-1})

(d) The electrocardiogram

Einthoven's triangle
Lead I: RA–LA
Lead II: RA–LL
Lead III: LA–LL

(e) Coronary circulation

Aorta
Right coronary artery
Left circumflex artery
Posterior descending artery
Marginal branch of right coronary artery
Pulmonary artery
Left main coronary artery
Diagonal branch of left anterior descending artery
Left circumflex marginal artery
Left anterior descending artery

The heart consists of four chambers—two thin-walled **atria** and two muscular **ventricles**. The atria are separated from the ventricles by a band of fibrous connective tissue (**annulus fibrosus**), which provides a skeleton for the attachment of muscle and the insertion of cardiac valves. It also prevents electrical conduction between the atria and ventricles, except at the **atrioventricular node** (AV node). The walls of the heart are formed from **cardiac muscle (myocardium)**. As the systemic circulation has a 10–15-fold greater resistance to flow than the pulmonary circulation, the left ventricle needs to develop more force and has more muscle than the right ventricle. The inner surface of the heart is covered by a thin layer of cells called the **endocardium**, similar to vascular **endothelial** cells (Chapter 17). This provides an anti-thrombogenic surface (inhibits clotting). The outer surface is covered by the **epicardium**, a layer of mesothelial cells. The whole heart is enclosed in a thin fibrous sheath (the **pericardium**), containing interstitial fluid as a lubricant, which protects the heart from damage caused by friction and prevents excessive enlargement.

Cardiac valves
Blood flows from the right atrium into the right ventricle via the **tricuspid** (three cusps or leaflets) AV valve, and from the left atrium to the left ventricle via the **mitral** (two cusps) AV valve. The AV valves are prevented from being everted into the atria by the high pressures developed in the ventricles by fine cords (**cordae tendinae** or *trabeculae*) attached between the edge of the valve cusps and **papillary muscles** in the ventricles (Fig. 13a). Blood is ejected from the right ventricle through the **pulmonary semilunar** valve into the pulmonary artery, and from the left ventricle via the **aortic semilunar** valve into the aorta; both semilunar valves have three cusps. The cusps are formed from connective tissue covered in a thin layer of **endocardial** or **endothelial** cells. When closed, the cusps form a tight seal at the **commissure** (line at which the edges meet). Both sets of valves open and close passively according to the pressure difference across them. Disease or the malformation of valves can have serious consequences. **Stenosis** describes narrowed valves; stenotic AV valves impair ventricular filling, and stenotic outflow valves increase **afterload** and thus ventricular work. **Incompetent** valves do not close properly and leak (**regurgitate**).

The cardiac pacemaker, conduction of the impulse and electrocardiogram
Cardiac muscle is described in Chapter 53. The heart beat is initiated in the **sinoatrial node** (SA node), a region of specialized myocytes in the right atrium, close to the coronary sinus. Spontaneous depolarization of the SA node (Chapter 15) provides the impulse for the heart to contract. Its rate is modulated by **autonomic nerves**. Action potentials (Chapter 15) in the SA node activate ad-jacent atrial myocytes via **gap junctions** contained within the **intercalated discs**; **desmosomes** provide a physical link (Fig. 13b & Chapter 15). A wave of depolarization and contraction therefore sweeps through the atrial muscle. This is prevented from reaching the ventricles directly by the **annulus fibrosus** (see above), and the impulse is channelled through the **AV node**, located between the right atrium and ventricle near the atrial septum.

The AV node contains small cells and conducts slowly; it therefore delays the impulse for ~120 ms, allowing time for atrial contraction to complete ventricular filling. Once complete, effective pumping requires rapid activation throughout the ventricles, and the impulse is transmitted from the AV node by specialized, wide and thus fast conducting myocytes in the **bundle of His** and **Purkinje fibres**, by which it is distributed over the inner surface of both ventricles (Fig. 13c). From here, a wave of depolarization and contraction moves from myocyte to myocyte across the endocardium until the whole ventricular mass is activated.

Electrocardiogram (Fig. 13d). The waves of depolarization through the heart cause *local currents* in surrounding fluid, which are detected at the body surface as small changes in voltage. This forms the basis of the **electrocardiogram (ECG)**. The classical ECG records the voltage between the left and right arm (**lead I**), the right arm and left leg (**lead II**), and the left arm and left leg (**lead III**). This is represented by **Einthoven's triangle** (Fig. 13d). The size of voltage at any time depends on the quantity of muscle depolarizing (more cells generate more current), and the **direction** in which the wave of depolarization is travelling (i.e. it is a **vector** quantity). Thus, lead II shows the largest deflection during ventricular depolarization, as the muscle mass is greatest and depolarization travels from apex to base, more or less parallel to a line from the left hip to the right shoulder. The basic interpretation of the ECG is described in Chapter 14.

The coronary circulation
The heart requires a rich blood supply, which is derived from the **left** and **right coronary arteries** arising from the aortic sinus (Fig. 13e). Cardiac muscle has an extensive system of capillaries. Most of the blood returns to the right atrium via the **coronary sinus**. The **large** and **small** coronary veins run parallel to the right coronary arteries, and empty into the coronary sinus. Small vessels, such as the **Thebesian veins**, empty into the cardiac chambers directly. The left ventricle is mostly supplied by the left coronary artery; occlusion in coronary artery disease can lead to serious damage. The coronary circulation is, however, capable of developing a good collateral system over time, where new arteries bypass occlusions and improve perfusion. During systole, contraction of the ventricles compresses the coronary arteries and suppresses blood flow; *more than 85% of left ventricular perfusion occurs during diastole*. This is problematical in disease if the heart rate is increased (e.g. exercise), as the diastolic interval is shorter.

14 The cardiac cycle

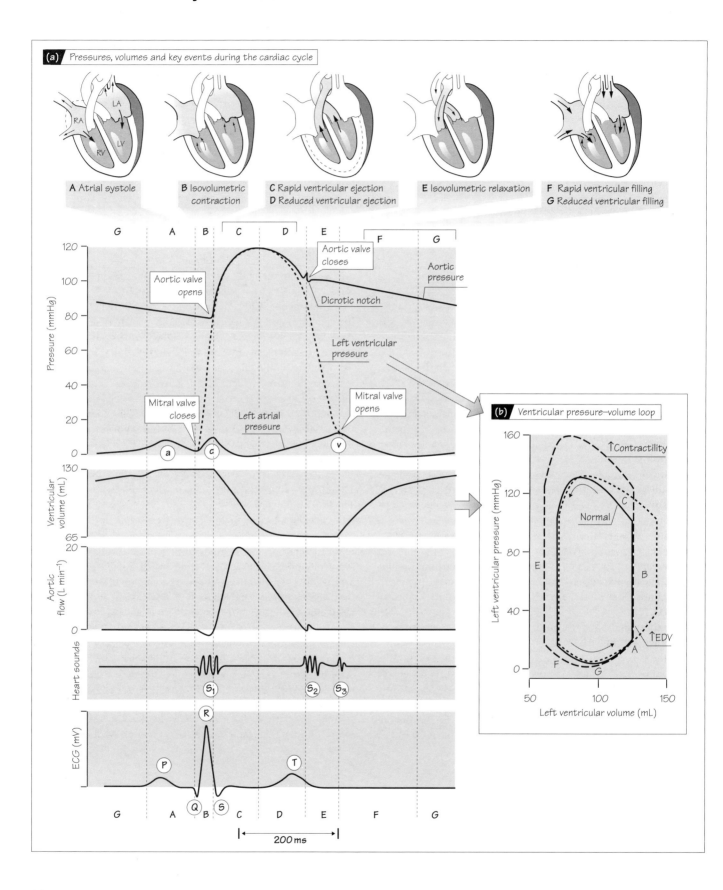

(a) Pressures, volumes and key events during the cardiac cycle

A Atrial systole

B Isovolumetric contraction

C Rapid ventricular ejection
D Reduced ventricular ejection

E Isovolumetric relaxation

F Rapid ventricular filling
G Reduced ventricular filling

Aortic valve opens

Aortic valve closes

Aortic pressure

Dicrotic notch

Left ventricular pressure

Mitral valve closes

Left atrial pressure

Mitral valve opens

Pressure (mmHg)

Ventricular volume (mL)

Aortic flow (L min⁻¹)

Heart sounds

ECG (mV)

200 ms

(b) Ventricular pressure–volume loop

↑Contractility

Normal

↑EDV

Left ventricular pressure (mmHg)

Left ventricular volume (mL)

The cardiac cycle (Fig. 14a) describes the events that occur during one beat of the heart. These are shown in the figure for the left side of the heart, together with the pressures and volumes in the chambers and main vessels. At the start of the cycle, towards the end of **diastole**, the whole of the heart is relaxed. The atrioventricular (AV) valves are open (right, **tricuspid**; left, **mitral**), because the atrial pressure is still slightly greater than the ventricular pressure. The **pulmonary** and **aortic valves** are closed, as the pulmonary artery and aortic pressures are greater than that in the ventricles. The cycle starts when the **sinoatrial node** (SA node) initiates atrial systole (Chapter 15).

Atrial systole (A). At rest, atrial contraction only contributes the last ~20% of final ventricular volume, as most of the filling has already occurred due to venous pressure. The atrial contribution increases with heart rate, as diastole shortens and there is less time for ventricular filling. There are no valves between the veins and atria, and atrial systole causes a small pressure rise in the great veins (the **a wave**). The ventricular volume after filling is complete (**end-diastolic volume**, EDV) is ~120–140 mL in humans. The **end-diastolic pressure** (EDP) is less than 10 mmHg, and is higher in the left ventricle than in the right due to the thicker and therefore stiffer left ventricular wall. EDV strongly affects the strength of ventricular contraction (see **Starling's law**; Chapter 16). Atrial depolarization causes the **P wave** of the electrocardiogram (**ECG**); it should be noted that atrial *repolarization* is too diffuse to be seen on the ECG.

Ventricular systole (B, C). The ventricular pressure rises sharply during contraction, and the AV valves close as soon as this is greater than the atrial pressure. This causes a vibration which is heard as the **first heart sound (S$_1$)**. Ventricular depolarization is associated with the **QRS** complex of the ECG. For a short period, whilst force is developing, both the AV and outflow (semilunar) valves are closed and there is no ejection, as the ventricular pressure is still less than that in the pulmonary artery and aorta. This is called **isovolumetric contraction (B)**, as the ventricular volume does not change. The increasing pressure makes the AV valves bulge into the atria, causing a small atrial pressure wave (**c wave**).

Ejection. Eventually, the ventricular pressure exceeds that in the aorta or pulmonary artery, the outflow valves open and blood is ejected. The flow is initially very rapid (**rapid ejection phase, C**) but, as contraction wanes, ejection is reduced (**reduced ejection phase, D**). During the second half of ejection, the ventricles stop actively contracting, and the muscle starts to repolarize; this is associated with the **T wave** of the ECG. The ventricular pressure during the reduced ejection phase is slightly less than that in the artery, but blood continues to flow out of the ventricle because of momentum. Eventually, the flow briefly reverses, causing the closure of the outflow valve, a small increase in aortic pressure (**dicrotic notch**) and the **second heart sound (S$_2$)**. The amount of blood ejected in one beat is the **stroke volume**, ~70 mL. About 50 mL of blood is therefore left in the ventricle at the end of systole (**end-systolic volume, ESV**). The proportion of EDV that is ejected (stroke volume/EDV) is the **ejection fraction**.

Diastole. Immediately following the closure of the outflow valves, the ventricles rapidly relax. The AV valves remain closed, however, because the ventricular pressure is initially still greater than that in the atria (**isovolumetric relaxation, E**). This is called isometric relaxation because again the ventricular volume does not change. Meanwhile, the atrial pressure has been increasing due to filling from the veins (**v wave**). When the ventricular pressure falls sufficiently, the AV valves open and the atrial pressure falls as the ventricles rapidly refill (**rapid filling phase, F**). This is assisted by elastic recoil of the ventricular walls, essentially sucking in the blood. Filling due to venous flow alone is slower during the last two-thirds of diastole (**reduced filling phase, G**). Diastole is twice the length of systole at rest, but decreases as the heart rate increases.

The ventricular pressure–volume loop
The ventricular pressure plotted against volume generates a loop (Fig. 14b), the area of which represents the work performed. Its shape is affected by the force of ventricular contraction (contractility), factors that alter refilling (**EDV**) and the pressure against which the ventricle has to pump (e.g. aortic pressure, **afterload**). An estimate of **stroke work** is calculated from the mean arterial pressure × stroke volume.

The pulse
The **peripheral arterial pulse** reflects the pressure waves travelling down through the blood from the heart; these move much faster than the blood itself. The shape of the pulse is affected by the compliance and diameter of the artery; stiff (e.g. atherosclerosis) or small arteries have sharper pulses because they cannot absorb the energy so easily. The **jugular venous pulse** reflects the right atrial pressure, as there is no valve between the jugular vein and right atria, and has corresponding **a**, **c**, and **v waves**.

Heart sounds
Heart sounds are caused by vibrations in the blood due, for example, to closure of the cardiac valves (see above). Normally, only the **first** and **second** heart sounds are detectable (S$_1$, S$_2$), although a third sound (S$_3$) can occasionally be heard in fit young people. When the atrial pressure is raised (e.g. in heart failure), both a third and fourth sound may be heard, associated with rapid filling and atrial systole, respectively; this sounds like a galloping horse (**gallop rhythm**). Cardiac **murmurs** are caused by turbulent blood, and a benign murmur is sometimes heard in young people during the ejection phase. Pathological murmurs are associated with the narrowing of valves (**stenosis**), or **regurgitation** of blood backwards through valves that do not close properly (**incompetence**).

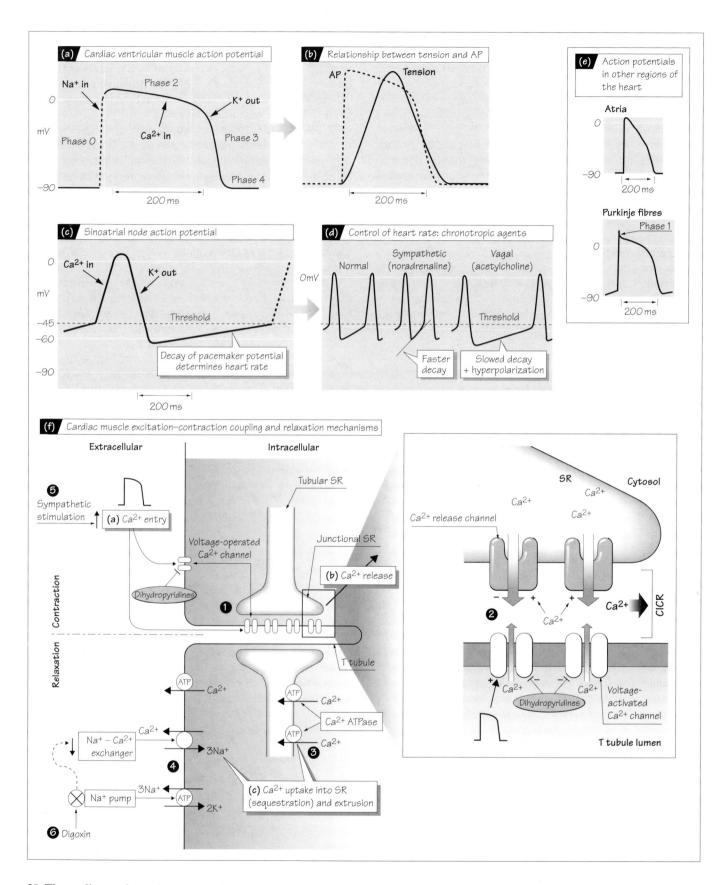

The process linking depolarization to contraction is called **excitation–contraction coupling**. The basics of action potentials (**APs**) are described in Chapter 6.

Cardiac muscle electrophysiology

Ventricular muscle action potential (Fig. 15a). The resting potential of ventricular myocytes is approximately -90 mV, and stable (phase 4). An AP is initiated when the myocyte is depolarized to a **threshold potential** of approximately -65 mV, as a result of transmission from an adjacent myocyte via **gap junctions** (Chapter 13). Fast, voltage-gated Na^+ channels are activated, leading to an inward current which depolarizes the membrane rapidly towards $+30$ mV. This initial depolarization or **upstroke** (phase 0; Fig. 15a) is similar to that in nerve and skeletal muscle, and assists transmission to the next myocyte. The Na^+ current rapidly inactivates, but, in cardiac myocytes, the initial depolarization activates voltage-gated Ca^{2+} channels (**L-type channels**; threshold approximately -40 mV), through which Ca^{2+} floods into the cell. The resultant inward current prevents the cell from repolarizing, and causes a **plateau phase** (phase 2) that is maintained for ~250 ms until the L-type channels inactivate. The cardiac AP is thus much longer than that in nerve or skeletal muscle (~300 ms vs. ~2 ms). Repolarization occurs due to an outward K^+ current (phase 3). The plateau and associated Ca^{2+} entry are essential for contraction; the blockade of L-type channels (e.g. **dihydropyridines**) reduces force. As the AP lasts almost as long as contraction (Fig. 15b), its *refractory period* (Chapter 6) prevents another AP being initiated until the muscle relaxes; thus, cardiac muscle cannot exhibit tetanus (see Chapter 52).

The sinoatrial node and origin of the heart beat

The sinoatrial node (SA node) AP differs from that in ventricular muscle (Fig. 15c). The resting potential starts at a more positive value (approximately -60 mV) and decays steadily with time until it reaches a threshold of around -40 mV, when an AP is initiated. The upstroke of the AP is **slow**, as it is not due to the activation of fast Na^+ channels, but instead slow **L-type Ca^{2+} channels**; the SA node contains no functional fast Na^+ channels. The slow upstroke means that conduction between SA nodal myocytes is slow; this is important in the **atrioventricular node** (**AV node**), which has a similar AP. The rate of decay of the SA node resting potential determines the time it takes to reach threshold and to generate another AP, and hence determines the heart rate; it is therefore called the **pacemaker potential**. The pacemaker potential decays because of a slowly reducing outward K^+ current set against inward currents. Factors that affect these currents alter the rate of decay and the time to reach threshold, and hence the heart rate, and are called **chronotropic agents**. The sympathetic transmitter, noradrenaline (norepinephrine), is a *positive chronotrope* that increases the rate of decay and thus the heart rate, whereas the parasympathetic transmitter, **acetylcholine**, lengthens the time to reach threshold and decreases the heart rate (Fig. 15d).

Action potentials elsewhere in the heart (Fig. 15e). Atria have a similar but more triangular AP to the ventricles. **Purkinje fibres** in the conduction system are also similar to ventricular myocytes, but have a spike (phase 1) at the peak of the upstroke, reflecting a larger Na^+ current that contributes to their fast conduction velocity (see Chapter 7). Other atrial cells, the AV node, bundle of His and Purkinje system may also exhibit decaying resting potentials that can act as pacemakers. However, the SA node is normally fastest and predominates. This is called **dominance** or **overdrive suppression**.

Excitation–contraction coupling (Fig. 15f)

Contraction. Cardiac muscle contracts when intracellular Ca^{2+} rises above 100 nM. Although Ca^{2+} entry during the AP is essential for contraction, it only accounts for ~25% of the rise in intracellular Ca^{2+}. The rest is released from Ca^{2+} stores in the **sarcoplasmic reticulum** (SR). APs travel down invaginations of the sarcolemma called **T-tubules**, which are close to, but do not touch, the **terminal cisternae** of the SR ❶. During the AP plateau, Ca^{2+} enters the cell and activates Ca^{2+}-sensitive **Ca^{2+} release channels** in the SR ❷, allowing stored Ca^{2+} to flood into the cytosol; this is **Ca^{2+}-induced Ca^{2+} release** (CICR). The amount of Ca^{2+} released depends on how much is stored and how much Ca^{2+} enters during the AP. Modulation of the latter is a key way in which cardiac muscle force is regulated (see below). Peak intracellular $[Ca^{2+}]$ normally rises to ~2 µM, although maximum contraction occurs above 10 µM.

Relaxation. Ca^{2+} is rapidly pumped back into the SR (**sequestered**) by adenosine triphosphate (ATP)-dependent Ca^{2+} pumps (Ca^{2+}-ATPase) ❸. However, Ca^{2+} that entered the myocyte during the AP must also be removed again. This is primarily performed by the **Na^+–Ca^{2+} exchanger** in the membrane, which pumps one Ca^{2+} ion out in exchange for three Na^+ ions, using the Na^+ electrochemical gradient as an energy source ❹. This is relatively slow, and continues during diastole. If the latter is shortened, i.e. when the heart rate rises, more Ca^{2+} is left inside the cell and the cardiac force increases. This is the **staircase** or Treppe effect.

Regulation of contractility: inotropic agents (Fig. 15f)

Sympathetic stimulation increases cardiac muscle **contractility** (Chapter 16) because it causes the release of noradrenaline, a **positive inotrope**. Noradrenaline binds to β_1-adrenoceptors on the membrane and causes increased Ca^{2+} entry via L-type Ca^{2+} channels during the AP ❺, and thus increases Ca^{2+} release from the SR (❷; see above). Noradrenaline also accelerates Ca^{2+} sequestration into the SR ❸. The contractility is also increased by slowing the removal of Ca^{2+} from the myocyte. **Cardiac glycosides** (e.g. *digoxin*) inhibit the Na^+ pump which removes Na^+ from the cell (Chapter 5) ❻. Intracellular $[Na^+]$ therefore increases and the Na^+ gradient across the membrane is reduced. This depresses Na^+–Ca^{2+} exchange ❹, which relies on the Na^+ gradient for its motive force, and Ca^{2+} is pumped out of the cell less rapidly. Consequently, more Ca^{2+} is available inside the myocyte for the next beat, and force increases. **Acidosis** (blood pH <7.3) is **negatively inotropic**, largely because H^+ competes for Ca^{2+}-binding sites.

16 Control of cardiac output and Starling's law of the heart

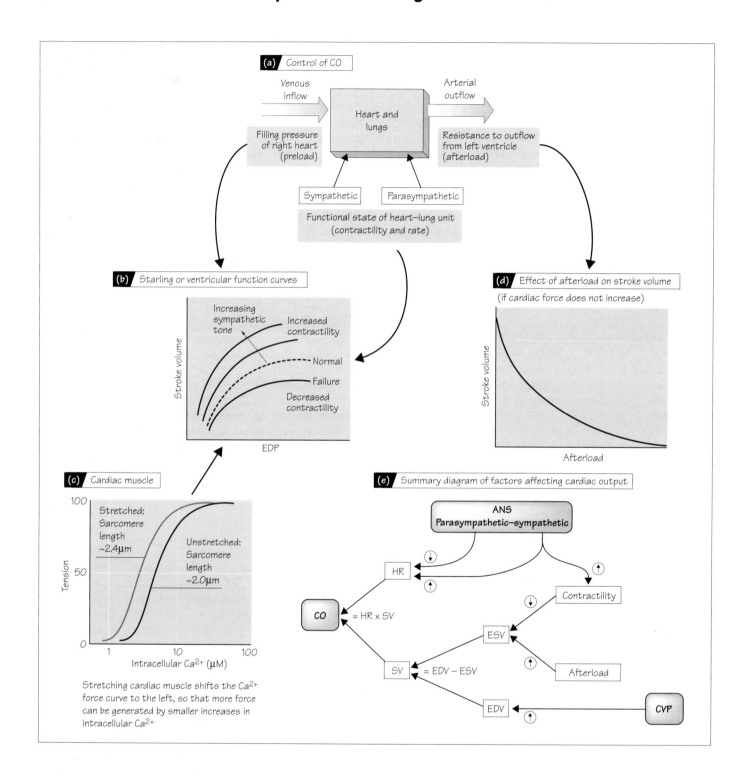

(a) Control of CO

Venous inflow → Heart and lungs → Arterial outflow

Filling pressure of right heart (preload)

Resistance to outflow from left ventricle (afterload)

Sympathetic   Parasympathetic

Functional state of heart–lung unit (contractility and rate)

(b) Starling or ventricular function curves

Increasing sympathetic tone
Increased contractility
Normal
Failure
Decreased contractility

Stroke volume (y-axis)
EDP (x-axis)

(d) Effect of afterload on stroke volume (if cardiac force does not increase)

Stroke volume (y-axis)
Afterload (x-axis)

(c) Cardiac muscle

Stretched: Sarcomere length ~2.4μm

Unstretched: Sarcomere length ~2.0μm

Tension (y-axis) 0, 50, 100
Intracellular Ca^{2+} (μM) (x-axis) 1, 10, 100

Stretching cardiac muscle shifts the Ca^{2+} force curve to the left, so that more force can be generated by smaller increases in intracellular Ca^{2+}

(e) Summary diagram of factors affecting cardiac output

ANS Parasympathetic–sympathetic

HR
Contractility
CO = HR × SV
ESV
SV = EDV − ESV
Afterload
EDV
CVP

Cardiac output is determined by the heart rate and stroke volume (cardiac output = heart rate × stroke volume). The stroke volume is influenced by the filling pressure (**preload**), the force developed by the cardiac muscle, and the pressure against which the heart has to pump (**afterload**). Both the heart rate and force development are modulated by the autonomic nervous system (Fig. 16a).

Filling pressure and Starling's law

The right ventricular end-diastolic pressure (**EDP**) is dependent on the central venous pressure (**CVP**); the left ventricular EDP is dependent on the pulmonary venous pressure. EDP and the compliance (stretchiness) of the ventricular wall determine the end-diastolic volume (**EDV**). A stiff ventricle, e.g. due to fibrosis after ischaemia, will not expand so readily and will have a smaller EDV for any EDP. As the EDV increases, the force of contraction developed during the subsequent systole also increases, with a consequent rise in the stroke volume. This is called the **Frank–Starling** relationship, and the graph relating stroke volume to EDP is called a **Starling** or **ventricular function curve** (Fig. 16b). The force of contraction is actually related to the degree of stretch of the cardiac muscle, and **Starling's law of the heart** is quoted as: '*The energy released during contraction depends on the initial fibre length*'.

This relationship between muscle length and force generation can be partly explained by the **sliding filament theory** of muscle contraction, as when muscle stretches more cross-bridges can form between myosin and actin (Chapter 50). The relationship is, however, steeper in cardiac than skeletal muscle. When cardiac muscle is stretched, there is also an increase in the Ca^{2+} **sensitivity**, so that more force is generated for any given rise in intracellular $[Ca^{2+}]$ (Fig. 16c). This mechanism involves troponin (Chapter 50), and gives greater sensitivity to small degrees of stretch.

Importance of Starling's law

The most important consequence of Starling's law is that *the stroke volumes of the left and right ventricles are matched*. If the output of the right ventricle was greater than that from the left ventricle, blood would accumulate in the lungs, pulmonary blood pressure would rise and fluid would be forced into the lung interstitium and alveoli (**pulmonary oedema**). This does not normally happen because any increase in pulmonary blood pressure increases the filling pressure, and hence EDV, of the left ventricle. The left ventricular stroke volume then increases according to Starling's law until it matches the right ventricular output again, at which point pulmonary blood pressure stops rising and a new equilibrium is reached. This also explains how an increase in CVP, which only directly affects the output of the right ventricle, causes an increase in cardiac output. Starling's law therefore also contributes to the rise in cardiac output during exercise when CVP may increase.

Postural hypotension. On standing from a prone position, gravity causes blood to pool in the legs and CVP falls. This, is turn, causes a fall in cardiac output (*due to Starling's law*) and, consequently, a fall in blood pressure. This **postural hypotension** is normally rapidly corrected by the **baroreceptor reflex** (Chapter 18), which causes venoconstriction (thus raising CVP), increases the heart rate and output, and restores the blood pressure. Even in healthy people, however, it can occasionally cause a temporary blackout (fainting or *syncope*) due to reduced cerebral perfusion. Fainting on parade is similar, when a lack of movement impairs **muscle pump** function (Chapter 12) with consequent pooling of blood.

The autonomic nervous system and contractility

The mechanisms underlying Starling's law are *intrinsic* to cardiac muscle. The autonomic nervous system provides an important *extrinsic* influence on cardiac output. **Sympathetic** stimulation and adrenaline (epinephrine) increase the heart rate and contractile force, whereas **parasympathetic** stimulation decreases the heart rate. Sympathetic stimulation causes the ventricular function curve to shift upwards and to the left, so that more force is generated at any given EDV (Fig. 16b). This increase in force without a change in length is called an increase in **contractility**. By definition, Starling's law does *not* cause an increase in contractility. The heart rate is modulated by **chronotropic** agents, and contractility by **inotropic** agents; the contractility is decreased in disease (e.g. *myocardial ischaemia*). The mechanisms underlying the modulation of heart rate and contractility are discussed in Chapter 15.

Afterload

Afterload is the load against which the heart has to work, and it should be intuitive that an increase in afterload will reduce output if the cardiac force cannot be increased (Fig. 16d). Afterload is normally related to aortic pressure for the left ventricle, and pulmonary artery pressure for the right. It therefore increases if blood pressure rises, or if there is stenosis (narrowing) of the outflow valves. The cardiac output can, however, usually be maintained as a consequence of Starling's law. When the afterload increases, there is initially a decline in the **ejection fraction** (proportion of EDV ejected per beat) and stroke volume. However, more blood is therefore left in the ventricle after systole, and also the outputs of the two sides of the heart no longer match. As a result, blood accumulates on the venous side and the filling pressure rises. The contractile force therefore increases according to Starling's law until it overcomes the increased afterload and, after a few beats, the cardiac output is restored. This mechanism may become insufficient in *heart failure*.

Control of cardiac output

Due to the compensation provided by Starling's law, cardiac output is actually only affected by the filling pressure of the right heart, i.e. CVP, and the effects of the autonomic nervous system on the heart rate and contractility (Fig. 16e). Cardiac output can thus be maintained in moderate heart disease, although at the expense of increased filling pressures. However, even if the stroke volume is normal, the ejection fraction is reduced because EDV has to be large to maintain this output. Thus, a **reduced ejection fraction** (<50%) and an **enlarged heart** are diagnostic for underlying heart disease.

17 Blood vessels

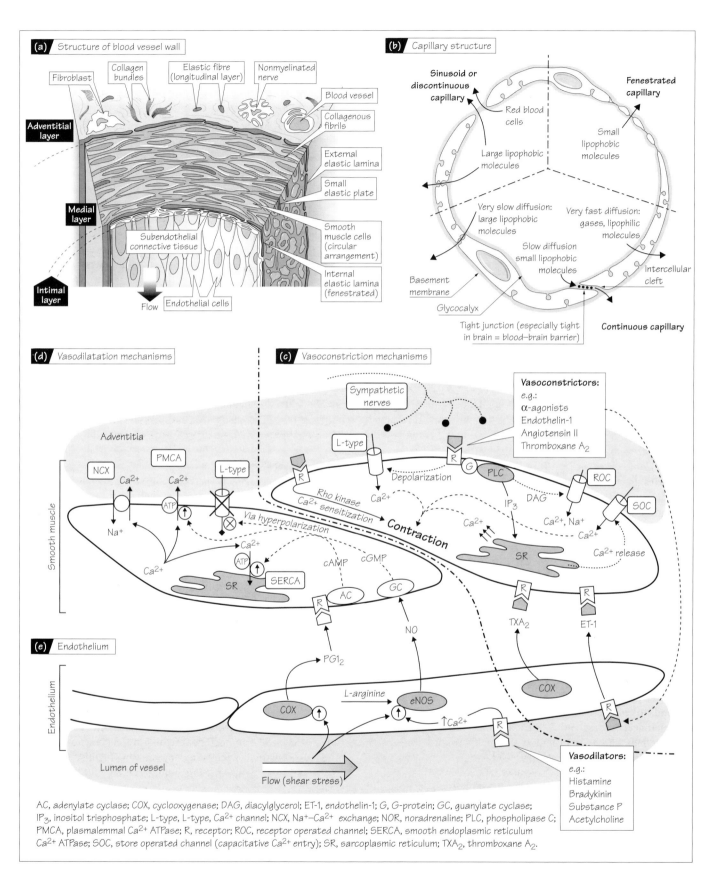

Structure

The walls of larger blood vessels comprise three layers: an inner **intima** (*tunica intima*) consisting of a thin layer of **endothelial cells**; a thick **media** (*tunica media*) containing **smooth muscle** and **elastin filaments** that provide elastic properties; and an outer **adventitia** (*tunica adventitia*) consisting of fibroblasts and nerves embedded in collagenous tissue (Fig. 17a). The layers are separated by inner and outer **elastic lamina**. In large vessels, the adventitia contains a network of blood vessels called the **vasa vasorum** (*vessel of vessels*) supplying the smooth muscle. Veins have a thinner media than arteries, and contain less smooth muscle. All three layers contain fibrous **collagen**, which acts as a framework to which cells are anchored.

Vascular smooth muscle cells are elongated, 15–100 μm in length, and tend to be orientated in a spiral around the vessel; the lumen therefore narrows as they contract. Cells are connected by **gap junctions**, allowing electrical coupling and depolarization to spread from cell to cell. The structure and function of smooth muscle are described in Chapter 53.

Capillaries and the smallest venules are formed from a single layer of endothelial cells supported on the outside by a 50–100 nm thick **basal lamina** containing collagen. The lumenal surface is covered by a glycoprotein network called the **glycocalyx**. There are three basic types of capillary, varying in permeability (Fig. 17b). **Continuous capillaries** have a low permeability, as junctions between the endothelial cells are very tight and prevent the diffusion of lipophobic molecules of >10 000 Da. They are found in skin, lungs, central nervous system and muscle. **Fenestrated capillaries** have less tight junctions and the endothelial cells are also punctured by 50–100 nm pores (**fenestrae**); they are therefore much more permeable. They are found where large amounts of fluid or material need to diffuse across the capillary wall, including endocrine glands, renal glomeruli and intestinal villa. **Discontinuous capillaries** are found in bone marrow, liver and spleen, and have gaps large enough for red blood cells to pass through. The microcirculation is discussed further in Chapter 19.

Regulation of function and smooth muscle excitation–contraction coupling

Vasoconstriction (Fig. 17c). Most vasoconstrictors bind to receptors and cause a guanosine triphosphate-binding protein (G-protein)-mediated elevation in intracellular $[Ca^{2+}]$, leading to contraction. Important vasoconstrictors include endothelin-1, angiotensin II (Chapter 31) and the sympathetic transmitter noradrenaline (norepinephrine) (Chapter 8).

Ca^{2+} release. Binding to a receptor activates **phospholipase C** which generates the second messengers inositol trisphosphate (**IP$_3$**) and diacylglycerol (**DAG**) from membrane phospholipids. IP$_3$ binds to receptors on the **sarcoplasmic reticulum** (SR) causing Ca^{2+} channels to open and Ca^{2+} to flood into the cytoplasm. This response may only be transient as the store rapidly empties, but may initiate *capacitative Ca^{2+} entry* (see below).

Ca^{2+} entry. Vasoconstrictors also cause depolarization, which activates Ca^{2+} entry via **L-type voltage-gated Ca^{2+} channels** as in cardiac muscle (Chapter 15). Unlike cardiac muscle, most types of vascular smooth muscle do not generate action potentials, but instead depolarization is graded, allowing graded entry of Ca^{2+}. **Receptor operated channels** may also be activated, some by DAG, through which both Ca^{2+} and Na^+ can enter the cell; the latter may contribute to depolarization. IP$_3$-stimulated emptying of Ca^{2+} stores can also directly activate **store operated channels** in the membrane, causing **capacitative Ca^{2+} entry**, although its importance may be limited to certain vascular beds (e.g. pulmonary).

An important modulatory pathway that enhances vasoconstriction involves **Ca^{2+} sensitization** of the contractile apparatus, i.e. more force for the same rise in $[Ca^{2+}]$. This is mediated by **Rho kinase**, although protein kinase C, which is activated by DAG, may also be involved. The relative importance of the above mechanisms depends on the vascular bed and vasoconstrictor. In small-resistance arteries, depolarization and voltage-gated Ca^{2+} entry are probably most important. Most systemic arteries exhibit a degree of **basal** (*myogenic*) **tone** in the absence of vasoconstrictors.

Ca^{2+} removal and vasodilatation (Fig. 17d). Ca^{2+} is pumped back into the SR (*sequestrated*) by the **smooth endoplasmic reticulum Ca^{2+}-ATPase** (**SERCA**) which can rapidly reduce cytosolic $[Ca^{2+}]$. Ca^{2+} is also removed from the cell by a **plasma membrane Ca^{2+}-ATPase** (PMCA) and **Na^+–Ca^{2+} exchange** (NCX; Chapter 15). Most endogenous vasodilators cause relaxation by increasing cyclic guanosine monophosphate (cGMP) (e.g. **nitric oxide**, NO) or cyclic adenosine monophosphate (cAMP) (e.g. **prostacyclin**, β-adrenergic receptor agonists). These second messengers act via protein kinase G (PKG) or protein kinase A (PKA), respectively. It is believed that both PKG and PKA lower intracellular $[Ca^{2+}]$, partly by stimulating SERCA and PMCA, and partly by hyperpolarizing the membrane (i.e. so voltage-gated Ca^{2+} entry is inhibited). L-type Ca^{2+} channel blocker drugs, such as **verapamil** or **dihydropyridines**, are clinically effective vasodilators.

The endothelium (Fig. 17e)

The endothelium plays a crucial role in the regulation of vascular tone. In response to substances in the blood or changes in blood flow, it can synthesize several important vasodilators, including **NO** (*endothelium-derived relaxing factor*, EDRF) and **prostacyclin** (prostaglandin I$_2$, PGI$_2$), as well as potent vasoconstrictors, such as **endothelin-1** and **thromboxane A$_2$** (TXA$_2$).

NO is synthesized by the endothelial **nitric oxide synthase** (eNOS) from L-arginine. eNOS activity and NO production are increased by factors that elevate intracellular $[Ca^{2+}]$, including local mediators such as **bradykinin**, **histamine** and **serotonin**, and some neurotransmitters (e.g. **substance P**). Increased flow (**shear stress**) also stimulates NO production, and additionally activates prostacyclin synthesis. The basal production of NO continuously modulates vascular resistance, as it has been found that inhibition of eNOS causes the blood pressure to rise. NO also inhibits platelet activation and **thrombosis** (inappropriate clotting) (Chapter 9).

Endothelin-1 is an extremely potent vasoconstrictor peptide which is released from the endothelium in the presence of many other vasoconstrictors, including angiotensin II, antidiuretic hormone (ADH; *vasopressin*) and noradrenaline, and may be increased in disease and hypoxia. As endothelin receptor blockade causes a fall in the peripheral resistance of healthy humans, it seems to contribute to the maintenance of blood pressure.

18 Control of blood pressure and blood volume

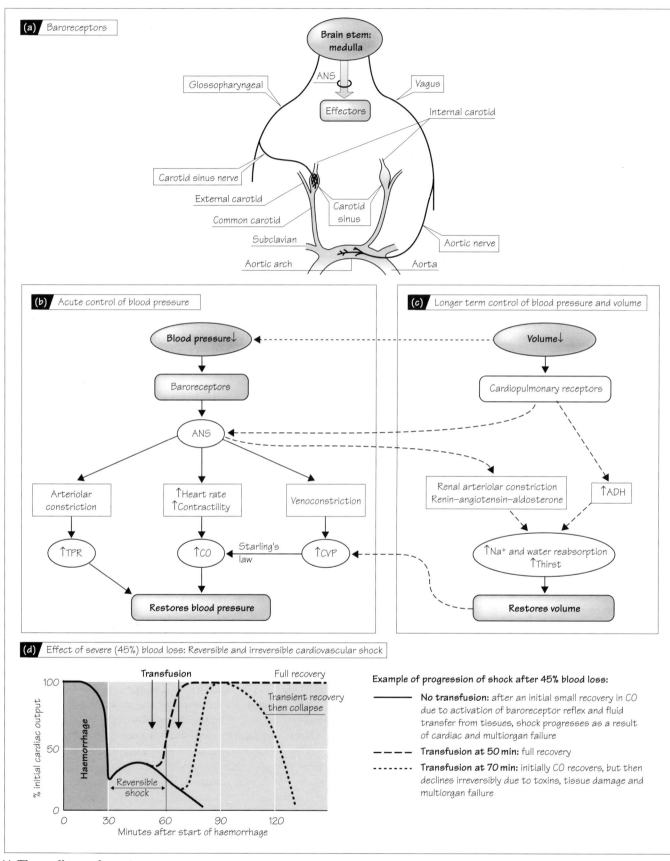

(a) Baroreceptors

Brain stem: medulla

ANS

Glossopharyngeal

Vagus

Effectors

Internal carotid

Carotid sinus nerve

External carotid

Common carotid

Carotid sinus

Subclavian

Aortic nerve

Aortic arch

Aorta

(b) Acute control of blood pressure

Blood pressure↓

Baroreceptors

ANS

Arteriolar constriction

↑Heart rate
↑Contractility

Venoconstriction

↑TPR

↑CO

Starling's law

↑CVP

Restores blood pressure

(c) Longer term control of blood pressure and volume

Volume↓

Cardiopulmonary receptors

Renal arteriolar constriction
Renin–angiotensin–aldosterone

↑ADH

↑Na⁺ and water reabsorption
↑Thirst

Restores volume

(d) Effect of severe (45%) blood loss: Reversible and irreversible cardiovascular shock

Transfusion

Full recovery

Transient recovery then collapse

100

% initial cardiac output

Haemorrhage

50

Reversible shock

0

0   30   60   90   120

Minutes after start of haemorrhage

Example of progression of shock after 45% blood loss:

—— **No transfusion:** after an initial small recovery in CO due to activation of baroreceptor reflex and fluid transfer from tissues, shock progresses as a result of cardiac and multiorgan failure

– – – **Transfusion at 50 min:** full recovery

······· **Transfusion at 70 min:** initially CO recovers, but then declines irreversibly due to toxins, tissue damage and multiorgan failure

Tissues can independently alter their blood flow by changing their vascular resistance. So that this does not have a knock-on effect elsewhere, the pressure head provided by the mean arterial blood pressure (MAP) must be controlled. MAP is determined by the **total peripheral resistance** (TPR) and **cardiac output** (MAP = cardiac output×TPR), which is itself dependent on the **central venous pressure** (CVP) (Chapter 16). CVP is highly dependent on the **blood volume**. Alterations of any of these variables may change MAP.

Effect of gravity. When standing, the blood pressure at the ankle is ~90 mmHg higher than that at the level of the heart, due to the weight of the column of blood between the two. Similarly, the pressure in the head is ~30 mmHg less than that at the level of the heart. Blood pressure is always measured at the level of the heart. Gravity does not affect the driving force between arteries and veins because arterial and venous pressures are affected equally.

Acute regulation of the mean arterial blood pressure: the baroreceptor reflex

Physiological regulation commonly involves **negative feedback**. This requires a **sensor** that detects the controlled variable (e.g. MAP), a **comparator** that compares the sensor output to a **set point**, and a **feedback pathway** driving **effectors** that adjust the variable until the difference between the sensor output and the set point is minimized (Chapter 2). The sensor for MAP is provided by **baroreceptors** (stretch receptors) located in the **carotid sinus** and **aortic arch** (Fig. 18a). A decrease in MAP reduces arterial wall stretch and *decreases* baroreceptor activity, resulting in decreased firing in afferent nerves travelling via the glossopharyngeal and vagus to the medulla of the brain stem, where the activity of the **autonomic nervous system** (Chapter 8) is coordinated. Sympathetic nervous activity *increases*, causing an increased heart rate and cardiac contractility (Chapter 16), peripheral vasoconstriction and an increase in TPR, and venoconstriction which increases CVP (Chapter 17). Parasympathetic activity *decreases*, contributing to the rise in heart rate (Chapter 15). MAP therefore returns to normal (Fig. 18b). An increase in MAP has the opposite effects.

The baroreceptors are most sensitive between 80 and 150 mmHg, and their sensitivity is increased by a large **pulse pressure** (Chapter 12). They also show **adaptation**; if a new pressure is maintained for a few hours, activity slowly returns towards (but not to) normal. The baroreceptor reflex is important for buffering *short-term* changes in MAP, for example when muscle blood flow increases rapidly in exercise. Cutting the baroreceptor nerves has a minor effect on average MAP, but fluctuations in pressure are much greater.

Posture. Changes in posture provide a good example of the acute baroreceptor reflex. When standing from a supine position, blood pools in the veins of the legs, causing a fall in CVP; cardiac output and MAP therefore fall (**postural hypotension**; Chapter 16). Baroreceptor firing is reduced and the baroreceptor reflex is activated. *Venoconstriction* reduces blood pooling and increases CVP which, with an *increase in heart rate and cardiac contractility*, returns cardiac output towards normal; peripheral *vasoconstriction* assists the restoration of MAP. The transient dizziness or blackout (**syncope**) occasionally experienced when rising rapidly is due to a

fall in cerebral perfusion that occurs before cardiac output and MAP can be corrected.

Long-term regulation: control of blood volume (Fig. 18c)

The blood volume is dependent on total body Na^+ and water. These are regulated by the kidneys and discussed in detail in Chapter 31; only a brief account is given here.

The activation of the baroreceptor reflex by a reduction in MAP leads to renal arteriolar constriction. This and the fall in MAP itself cause a reduction in renal perfusion pressure, which inhibits the excretion of Na^+ and water in the urine. Sympathetic stimulation and reduced arteriolar pressure also activate the **renin–angiotensin system** (Chapter 31) and the production of **angiotensin II**, a potent vasoconstrictor that increases TPR. Angiotensin II also stimulates the production of **aldosterone** from the adrenal cortex, which promotes renal Na^+ reabsorption. The net effect is Na^+ and water retention, and an increase in blood volume (Fig. 18d). Conversely, a rise in MAP increases Na^+ and water excretion.

Changes in blood volume are sensed directly by **cardiopulmonary receptors**; **veno-atrial receptors** are located around the join between the veins and atria, and **atrial receptors** in the atrial wall. These effectively respond to changes in CVP and *blood volume*. Stimulation (stretch) suppresses the renin–angiotensin system and sympathetic activity, and the secretion of **antidiuretic hormone** (ADH, vasopressin), which promotes renal water reabsorption (Chapters 30 & 31). The cardiopulmonary receptors normally cause **tonic depression**—cutting their efferent nerves increases the heart rate and causes vasoconstriction in the gut, kidney and skeletal muscle, thus raising MAP.

Cardiovascular shock and haemorrhage

Cardiovascular shock. This is an acute condition with inadequate blood flow throughout the body, commonly associated with a fall in MAP. It can result from reduced blood volume (**hypovolumic shock**), profound vasodilatation (**low-resistance shock**) or acute failure of the heart to pump (**cardiogenic shock**). The most common cause of hypovolumic shock is **haemorrhage**; others include severe burns, diarrhoea and vomiting (e.g. cholera). Low-resistance shock is due to the profound vasodilatation caused by bacterial infection (**septic shock**) or powerful allergic reactions (e.g. to bee stings or peanuts; **anaphylactic shock**).

Haemorrhage. Twenty per cent of the blood volume can be lost without significant problems, as the baroreceptor reflex mobilizes blood from capacitance vessels and maintains MAP. Volume is restored within 24 h because arteriolar constriction reduces the capillary pressure and fluid moves from tissues into the plasma (Chapter 19), urine production is suppressed (see above) and ADH and angiotensin II stimulate thirst. Greater loss (30–50%) can be survived, but only with transfusion within ~1 h (the '**golden hour**') (Fig. 18d). After this, **irreversible shock** generally develops, which is irretrievable even with transfusion. This is because the reduced MAP and consequent profound peripheral vasoconstriction cause tissue ischaemia and the build-up of toxins and acidity, which damage the microvasculature and heart and lead to *multiorgan failure*.

19 The microcirculation, filtration and lymphatics

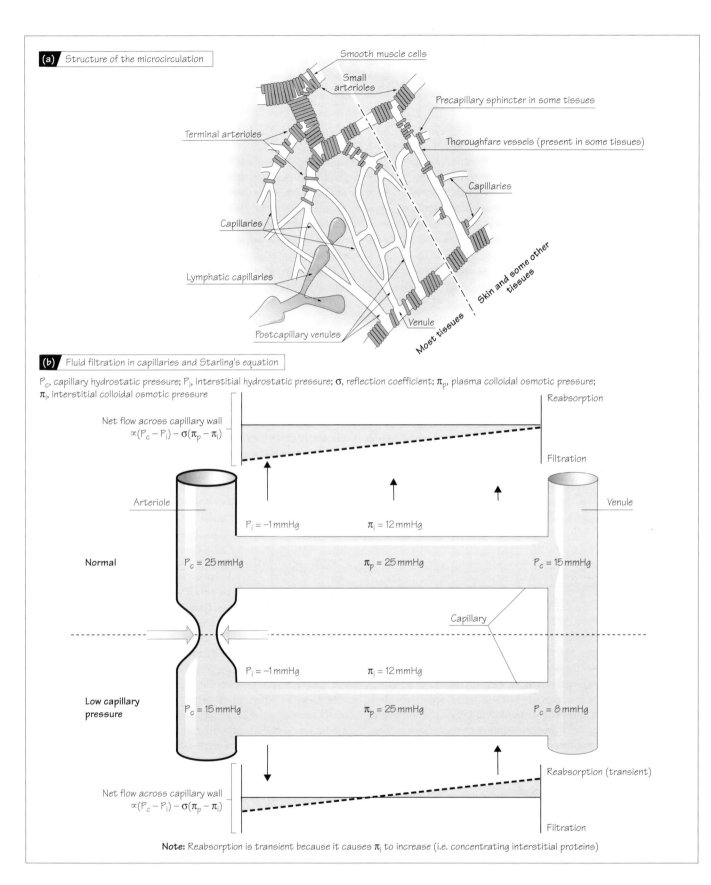

(a) Structure of the microcirculation

Smooth muscle cells

Small arterioles

Precapillary sphincter in some tissues

Terminal arterioles

Thoroughfare vessels (present in some tissues)

Capillaries

Capillaries

Lymphatic capillaries

Postcapillary venules

Venule

Most tissues

Skin and some other tissues

(b) Fluid filtration in capillaries and Starling's equation

P_c, capillary hydrostatic pressure; P_i, interstitial hydrostatic pressure; σ, reflection coefficient; π_p, plasma colloidal osmotic pressure; π_i, interstitial colloidal osmotic pressure

Reabsorption

Net flow across capillary wall
$\propto (P_c - P_i) - \sigma(\pi_p - \pi_i)$

Filtration

Arteriole

Venule

$P_i = -1\,\mathrm{mmHg}$

$\pi_i = 12\,\mathrm{mmHg}$

Normal

$P_c = 25\,\mathrm{mmHg}$

$\pi_p = 25\,\mathrm{mmHg}$

$P_c = 15\,\mathrm{mmHg}$

Capillary

$P_i = -1\,\mathrm{mmHg}$

$\pi_i = 12\,\mathrm{mmHg}$

Low capillary pressure

$P_c = 15\,\mathrm{mmHg}$

$\pi_p = 25\,\mathrm{mmHg}$

$P_c = 8\,\mathrm{mmHg}$

Reabsorption (transient)

Net flow across capillary wall
$\propto (P_c - P_i) - \sigma(\pi_p - \pi_i)$

Filtration

Note: Reabsorption is transient because it causes π_i to increase (i.e. concentrating interstitial proteins)

The **microcirculation** is perhaps the *raison d'étre* for the cardio-vascular system, as it is here that exchange between blood and tissues occurs. It consists of the smallest (**terminal**) arterioles and the **exchange vessels**—capillaries and **small venules** (Chapter 12). Blood flow into the microcirculation is regulated by the vasoconstriction of small arterioles, activated by sympathetic stimulation through numerous nerve endings in their walls (Chapters 8 & 18). Each small arteriole feeds many capillaries via several **terminal arterioles** (Fig. 19a), which are not innervated. Instead, the vasoconstriction of terminal arterioles is mediated by **local metabolic products** (Chapter 20), allowing perfusion to be matched to metabolism. A *few* tissues (e.g. mesenteric, skin) have **thoroughfare vessels** connecting small arterioles and venules directly. These contain a few smooth muscle cells spaced along their length (Fig. 19a). Between these vessels and capillaries, narrow rings of smooth muscle (**precapillary sphincters**) regulate capillary flow; these are not present in most other tissues.

Transcapillary exchange

Water, gases and other substances cross the capillary wall mainly by **diffusion** down their concentration gradients (Chapter 11). O_2 and CO_2 are highly **lipophilic** (soluble in lipids), and can cross the endothelial lipid bilayer membrane easily. This is, however, impermeable to **hydrophilic** ('water-loving', lipid-insoluble) molecules, such as glucose, and **polar** (charged) molecules and ions (electrolytes). Such substances mainly cross the wall of **continuous capillaries** through the gaps between endothelial cells. This is slowed by **tight junctions** between cells and by the **glycocalyx** (Chapter 17), so that diffusion is 1000–10 000 times slower than for lipophilic substances. This **small pore** system also prevents the diffusion of substances greater than 10 000 Da (e.g. plasma proteins). The latter can cross the capillary wall, but extremely slowly; this may involve **large pores** through endothelial cells. **Fenestrated capillaries** (gut, joints, kidneys) are 10-fold more permeable than continuous capillaries because of pores called **fenestrae** (from the Latin for '*windows*'), whereas **discontinuous capillaries** are highly permeable due to large spaces between endothelial cells, and occur where red cells need to cross the capillary wall (bone marrow, spleen, liver) (Chapter 17).

Filtration (Fig. 19b)

The capillary walls are much more permeable to water and electrolytes than to proteins (see above). The concentration of electrolytes (e.g. Na^+, Cl^-), and therefore the osmotic pressure exerted by them (**crystalloid osmotic pressure**), is very similar in plasma and interstitial fluid, and has little effect on fluid movement. The protein concentration in plasma is, however, greater than that in interstitial fluid, and the component of osmotic pressure exerted by proteins (**colloidal osmotic** or **oncotic pressure**) is therefore also greater. Water tends to flow from a *low* to a *high* osmotic pressure, but from a *high* to a *low* hydrostatic pressure. The net flow of water across the capillary wall is therefore determined by the balance between the hydrostatic (P) and colloidal osmotic (π) pressures, according to **Starling's equation, flow $\propto (P_c - P_i) - \sigma(\pi_p - \pi_i)$**,

where $(P_c - P_i)$ is the difference in hydrostatic pressure between capillary and interstitial fluid, and $(\pi_p - \pi_i)$ is the difference in colloidal osmotic pressure between plasma and interstitial fluid; $(\pi_p - \pi_i)$ has an average value of ~13 mmHg. σ is the **reflection coefficient** (~0.9), a measure of how difficult it is for plasma proteins to cross the capillary wall.

The capillary hydrostatic pressure normally varies from ~25 mmHg at the arteriolar end to ~15 mmHg at the venous end, whereas the interstitial hydrostatic pressure is approximately –1 mmHg. $(P_c - P_i)$ is therefore greater than $\sigma(\pi_p - \pi_i)$ along the length of the capillary, resulting in the **net filtration** of water into the interstitial space (Fig. 19b). Although arteriolar constriction will reduce capillary pressure and therefore lead to the reabsorption of fluid, this will normally be transient due to the concentration of interstitial fluid (i.e. increased π_i). A reduction in plasma protein (e.g. *starvation*), or a loss of endothelium integrity and thus diffusion of protein into the interstitial space (e.g. *severe inflammation, ischaemia*), will similarly reduce $(\pi_p - \pi_i)$, leading to enhanced filtration and loss of fluid into the tissues. This is also caused by a high venous pressure (**oedema**; see below).

Lymphatics

Fluid filtered by the microcirculation (~8 L per day) is returned to the blood by the **lymphatic system**. Lymphatic capillaries are blind-ended bulbous tubes (diameter, ~15–75 μm) walled with endothelial cells (Fig. 19a). These allow the entry of fluid, proteins and bacteria, but prevent their exit. Lymphatic capillaries merge into **collecting lymphatics** and then larger lymphatic vessels, both containing smooth muscle and **unidirectional valves**. Lymph is propelled, by smooth muscle constriction and compression of the vessels by body movement, into **afferent lymphatics** and then the **lymphatic nodes**, where bacteria and other foreign materials are removed by phagocytes. Most fluid is reabsorbed here by capillaries, with the remainder returning via **efferent lymphatics** and the thoracic duct into the subclavian veins. Lymphatics are also important for lipid absorption in the gut.

Oedema

Oedema is swelling of the tissues due to excess fluid in the interstitial space. It is caused when filtration is increased to the extent that the lymphatics are not able to remove the fluid fast enough (see above), or by dysfunctional lymphatic drainage (e.g. *elephantiasis*, the blockage of lymphatics with filarial nematode worms). Reduced venous drainage (increased venous pressure) also increases filtration and can lead to oedema; standing without moving the legs prevents the operation of the **muscle pump** (Chapter 12), local venous pressure rises, and the legs swell. In **congestive heart failure**, reduced cardiac function results in increased pulmonary and central venous pressure (Chapter 16), leading, respectively, to **pulmonary oedema** (alveoli fill with fluid) and **peripheral oedema** (swelling of the legs and liver, and accumulation of fluid in the peritoneum (*ascites*)). Severe protein starvation can cause generalized oedema and a grossly swollen abdomen due to ascites and an enlarged liver (*kwashiorkor*).

20 Local control of blood flow and special circulations

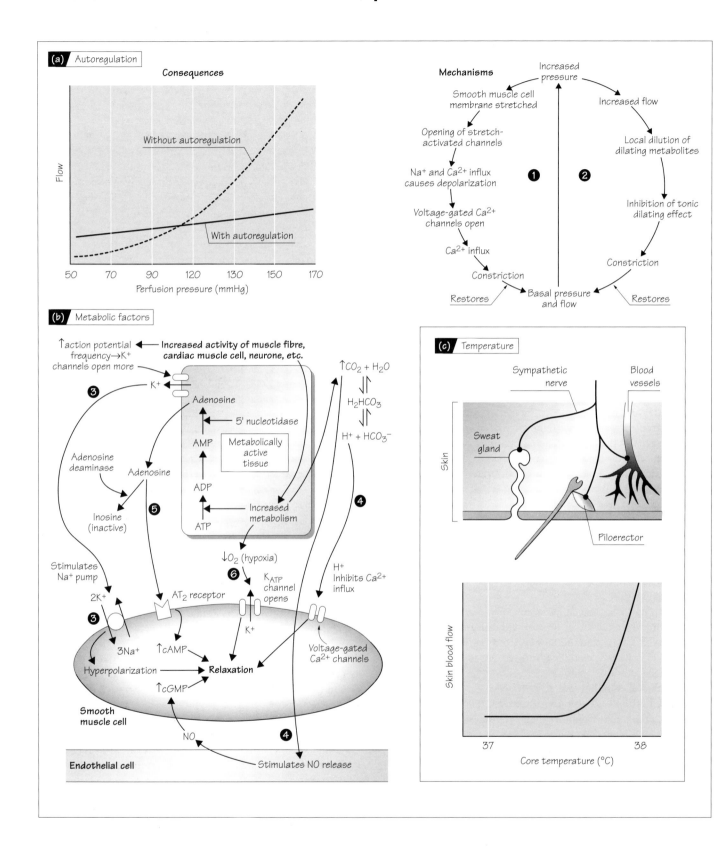

Local control of blood flow

In addition to the central control of blood pressure and cardiac output, tissues need to be able to regulate their own blood flow to match their requirements. This is provided by **autoregulation**, **metabolic factors** and **autocoids** (local hormones).

Autoregulation (Fig. 20a). Autoregulation is the ability to maintain a constant flow in the face of variations in pressure between ~50 and 170 mmHg. It is particularly important in the brain, kidney and heart. Two mechanisms contribute to autoregulation. The **myogenic response** ❶ involves arteriolar constriction in response to stretching of the vessel wall, probably due to the activation of smooth muscle **stretch-activated Ca^{2+} channels** and Ca^{2+} entry. A reduction in pressure and stretch closes these channels, causing vasorelaxation. The second mechanism is due to locally produced **vasodilating factors** ❷. An increase in blood flow dilutes these factors, causing vasoconstriction, whereas a decrease in blood flow allows accumulation, causing vasodilatation.

Metabolic factors (Fig. 20b). Many factors may contribute to **metabolic hyperaemia** (*increased blood flow*). The most important are K^+, CO_2 and **adenosine**, and, in some tissues, **hypoxia**. K^+ ❸ is released from active tissues and in ischaemia; local concentrations can increase to >10 mM. It causes relaxation, partly by stimulating the **Na^+ pump**, thus both increasing Ca^{2+} removal by the Na^+–Ca^{2+} exchanger and hyperpolarizing the cell (Chapter 17). The vasodilatory effects of increased CO_2 (**hypercapnia**) and **acidosis** ❹ are mediated largely through increased **nitric oxide** production (Chapter 17) and inhibition of smooth muscle Ca^{2+} entry. **Adenosine** ❺ is a potent vasodilator released from heart, skeletal muscle and brain during increased metabolism and hypoxia. It is produced from adenosine monophosphate (AMP), a breakdown product of adenosine triphosphate (ATP), and acts by stimulating the production of cyclic AMP (**cAMP**) in smooth muscle (Chapter 17). **Hypoxia** may reduce ATP sufficiently for K_{ATP} channels to activate ❻, causing hyperpolarization.

Autocoids are mostly important in special circumstances; two examples are given. In **inflammation**, mediators such as **histamine** and **bradykinin** cause vasodilatation and increase the permeability of exchange vessels, leading to swelling, but allowing access by leucocytes and antibodies to damaged tissues. The **activation of platelets** during clotting releases the vasoconstrictors **serotonin** and **thromboxane A_2**, so reducing blood loss (Chapter 9).

Special circulations

Skeletal muscle. This comprises ~50% of the body weight and, at rest, takes 15–20% of cardiac output; during exercise, this can rise to >80%. Skeletal muscle provides a major contribution to the **total peripheral resistance**, and sympathetic regulation of muscle blood flow is important in the **baroreceptor reflex**. At rest, most capillaries are not perfused, as their arterioles are constricted. Capillaries are **recruited** during exercise by **metabolic hyperaemia**, caused by the release of K^+ and CO_2 from the muscle, and **adenosine**. This *overrides* sympathetic vasoconstriction in working muscle; the latter reduces flow in non-working muscle, *conserving* cardiac output. It should be noted that muscular contraction compresses blood vessels and inhibits flow; in rhythmic (**phasic**) activity, metabolic hyperaemia compensates by vastly increasing flow during the relaxation phase. In **isometric** (static) contractions, reduced flow can cause **muscle fatigue**.

Brain. The occlusion of blood flow to the brain causes unconsciousness within minutes. The brain receives ~15% of cardiac output, and has a high capillary density. The endothelial cells of these capillaries have very tight junctions, and contain membrane transporters that control the movement of substances, such as ions, glucose and amino acids, and tightly regulate the composition of the cerebrospinal fluid. This arrangement is called the **blood–brain barrier**, and is continuous except where substances need to be absorbed or released from the blood (e.g. pituitary gland, choroid plexus). It can cause problems for the delivery of drugs to the brain, particularly antibiotics. The **autoregulation** of cerebral blood flow is highly developed, maintaining a constant flow for blood pressures between 50 and 170 mmHg. CO_2 and K^+ are particularly important metabolic regulators in the brain, increases causing a **functional hyperaemia** linking blood flow to activity. Hyperventilation reduces blood P_{CO_2}, and can cause fainting due to cerebral vasoconstriction.

Coronary circulation. The heart has a high metabolic demand and a dense capillary network. It is capable of extracting an unusually high proportion of oxygen from the blood (~70%). In exercise, the reduced diastolic interval (Chapter 14) and increased oxygen consumption demand a greatly increased blood flow, which is achieved under the influence of **adenosine**, K^+ and **hypoxia**. The heart therefore controls its own blood flow by a well-developed **metabolic hyperaemia**. This *overrides* the vasoconstriction mediated by sympathetic nerves (Chapters 8 & 17), and is assisted by circulating adrenaline (epinephrine) which causes vasodilatation via β_2-adrenergic receptors.

Skin (Fig. 20c). The main function of the cutaneous circulation is **thermoregulation**. Thoroughfare vessels (Chapter 19), formed from coiled **arteriovenous anastomoses** (AVAs), directly link arterioles and venules, allowing a high blood flow into the **venous plexus** and the radiation of heat. AVAs are found mostly in the hands, feet and areas of the face. Temperature is sensed in the **hypothalamus**. When low, sympathetic stimulation causes the vasoconstriction of cutaneous vessels; this also occurs following activation of the *baroreceptor reflex* by low blood pressure (e.g. *pale skin* in haemorrhage and shock) (Chapter 18). **Piloerection** (raising of skin hair, 'goosebumps') traps insulating air. Increased temperatures reduce sympathetic *adrenergic* stimulation, causing vasodilatation, whereas activation of **sympathetic *cholinergic* fibres** promotes **sweating** and the release of **bradykinin** which also causes vasodilatation. The net increase in blood flow may be 30-fold.

Pulmonary circulation. The pulmonary circulation is not controlled by either autonomic nerves or metabolic products, and the most important mechanism regulating flow is **hypoxic pulmonary vasoconstriction**, in which small arteries *constrict* to hypoxia. This is unique to the lung; hypoxia causes vasodilatation in systemic arteries (see above). Hypoxic pulmonary vasoconstriction diverts blood away from poorly ventilated areas of the lung, thus maintaining optimal **ventilation–perfusion matching** (Chapter 26); conversely, *global hypoxia* due to lung disease or altitude detrimentally increases the pulmonary artery pressure (*pulmonary hypertension*). The pulmonary capillary pressure is normally low (~7 mmHg), but fluid filtration still occurs because the interstitial pressure is low (approximately –4 mmHg) and the colloidal osmotic pressure is high (18 mmHg) (Chapter 19).

21 Introduction to the respiratory system

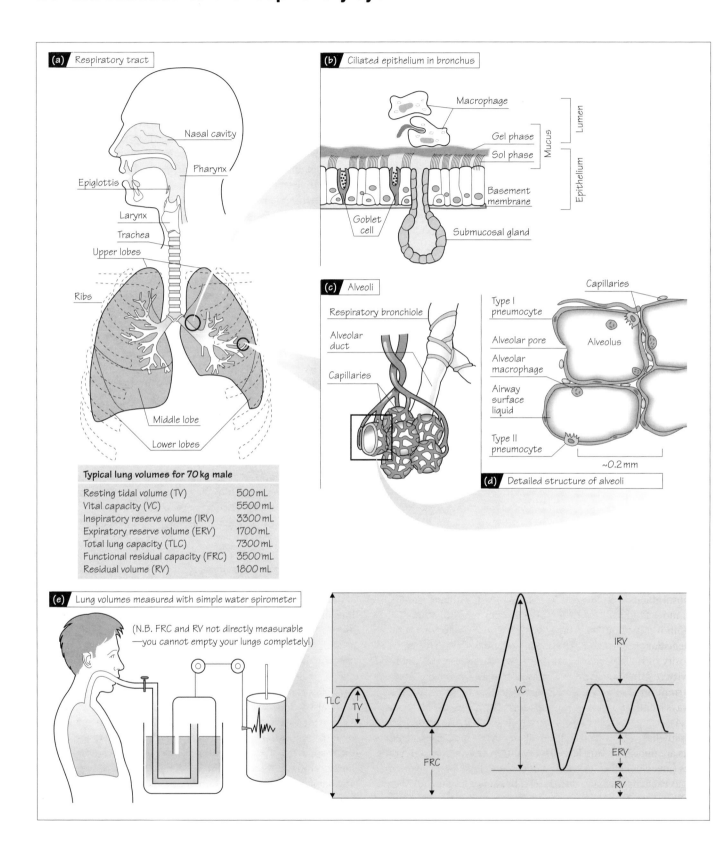

(a) Respiratory tract

Nasal cavity

Pharynx

Epiglottis

Larynx

Trachea

Upper lobes

Ribs

Middle lobe

Lower lobes

(b) Ciliated epithelium in bronchus

Macrophage

Gel phase

Sol phase

Mucus

Lumen

Epithelium

Basement membrane

Goblet cell

Submucosal gland

(c) Alveoli

Respiratory bronchiole

Alveolar duct

Capillaries

Capillaries

Type I pneumocyte

Alveolar pore

Alveolus

Alveolar macrophage

Airway surface liquid

Type II pneumocyte

~0.2 mm

(d) Detailed structure of alveoli

Typical lung volumes for 70 kg male	
Resting tidal volume (TV)	500 mL
Vital capacity (VC)	5500 mL
Inspiratory reserve volume (IRV)	3300 mL
Expiratory reserve volume (ERV)	1700 mL
Total lung capacity (TLC)	7300 mL
Functional residual capacity (FRC)	3500 mL
Residual volume (RV)	1800 mL

(e) Lung volumes measured with simple water spirometer

(N.B. FRC and RV not directly measurable
—you cannot empty your lungs completely!)

IRV

VC

TLC

TV

FRC

ERV

RV

The **upper respiratory tract** includes the nose, pharynx and larynx; the **lower respiratory tract** starts at the trachea (Fig. 21a). The two **lungs** are enclosed within the **thoracic cage**, formed from the ribs, sternum and vertebral column, with the dome-shaped **diaphragm** separating the thorax from the abdomen. The left lung has two lobes, the right three. The airways, blood vessels and lymphatics enter each lung at the lung root or **hilum**, where the **pulmonary nerve plexus** receives autonomic nerves from the vagus and sympathetic trunk. The vagus contains sensory afferents from **lung receptors** (Chapter 25) and bronchoconstrictor parasympathetic efferents leading to the airways; sympathetic nerves are *bronchodilatory* (Chapter 8). Each lung lobe is made up of several wedge-shaped **bronchopulmonary segments** supplied by their own segmental bronchus, artery and vein. The lungs are covered by a thin membrane (**visceral pleura**), continuous with the **parietal pleura** that lines the inside surface of the thoracic cage. The tiny space between the pleura is filled with lubricating **pleural fluid**.

Airways (Fig. 21a)

The **trachea** divides into two **main bronchi**; their walls contain U-shaped cartilage segments linked by smooth muscle. On entering the lung, the bronchi divide repeatedly into lobar, segmental (generation 3 and 4) and small (generation 5–11) bronchi, the smallest having a diameter of ~1 mm. These all have irregular cartilaginous plates and helical bands of smooth muscle. **Bronchioles** (generation 12–16) lack cartilage and are held open by surrounding lung tissue. The smallest (**terminal**) bronchioles lead to **respiratory bronchioles** (generation 17–19), and thence to **alveolar ducts** and **sacs** (generation 23), the walls of which form **alveoli** and contain only epithelial cells (Fig. 21c,d). Small pores (**alveolar pores**, *pores of Kohn*) allow pressure equalization between alveoli. The lungs contain ~17 million branches and ~300 million alveoli, providing an exchange surface of ~85 m^2. The **bronchial circulation** supplies airways down to the terminal bronchioles; respiratory bronchioles and below obtain nutrients from the **pulmonary circulation** (Chapter 12).

Epithelium and airway clearance

The airways from the trachea to the respiratory bronchioles are lined with **ciliated columnar epithelial cells**. **Goblet cells** and **submucosal glands** secrete a 10–15 μm thick, gel-like **mucus** that floats on a more fluid *sol phase* (Fig. 21b). Synchronous beating of the cilia moves the mucus and associated debris to the mouth (**mucociliary clearance**). Factors that increase the thickness or viscosity of the mucus (e.g. *asthma, cystic fibrosis*) or reduce cilia activity (e.g. *smoking*) impair mucociliary clearance and lead to recurrent infections. Mucus contains substances that protect the airways from pathogens (e.g. *antitrypsins, lysozyme, immunoglobulin A*).

Epithelial cells forming the walls of the alveoli and alveolar ducts are unciliated, and largely very thin **type I alveolar pneumocytes** (alveolar cells; *squamous epithelium*) (Fig. 21d). These form the gas exchange surface with the capillary endothelium (**alveolar-capillary membrane**). A few **type II pneumocytes** secrete **surfactant** which reduces the surface tension and prevents alveolar collapse (Chapter 22). **Macrophages** (*mobile phagocytes*) in the airways ingest foreign materials and destroy bacteria; in the alveoli, they take the place of cilia by clearing debris.

Respiratory muscles

The main respiratory muscles are inspiratory, the most important being the **diaphragm**; contraction pulls down the dome, reducing pressure in the thoracic cavity, and thus drawing air into the lungs. The **external intercostal muscles** assist by elevating the ribs and increasing the dimensions of the thoracic cavity. Quiet breathing is normally diaphragmatic; **accessory inspiratory muscles** (e.g. *scalene, sternomastoids*) aid inspiration if airway resistance or ventilation is high. Expiration is achieved by *passive recoil* of the lungs and chest wall, but, at high ventilation rates, this is assisted by the contraction of **abdominal muscles** which speed recoil of the diaphragm by raising abdominal pressure (e.g. exercise).

Lung volumes and pressures (Fig. 21e)

The **tidal volume** is the volume of air drawn into and out of the lungs during normal breathing; the **resting tidal volume** is normally ~ 500 mL but, like all lung volumes, is dependent on age, sex and height. The **vital capacity** is the maximum tidal volume, when an individual breathes in and out as far as possible. The difference in volume between a resting and maximum expiration is the **expiratory reserve volume**; the equivalent for inspiration is the **inspiratory reserve volume**. The volume in the lungs after a maximum inspiration is the **total lung capacity**, whilst that after a maximum expiration is the **residual volume**.

The **functional residual capacity** (FRC) is the volume of the lungs at the end of a normal breath, when the respiratory muscles are relaxed. It is determined by the balance between *outward elastic recoil* of the chest wall and *inward elastic recoil* of the lungs. These are coupled by the fluid in the small pleural space, which therefore has a negative pressure (**intrapleural pressure**: –0.2 to –0.5 kPa). Perforation of the chest therefore allows air to be sucked into the pleural space, and the chest wall expands whilst the lung collapses (**pneumothorax**). Diseases that affect lung elastic recoil alter FRC; *fibrosis* increases recoil and therefore reduces FRC, whereas in *emphysema*, where lung structure is lost, recoil is reduced and FRC increases.

During **inspiration**, the expansion of the thoracic cavity makes the intrapleural pressure more negative, causing the lungs and alveoli to expand, and reducing the alveolar pressure. This creates a pressure gradient between the alveoli and the mouth, drawing air into the lungs. During **expiration**, intrapleural and alveolar pressures rise, although, except during forced expiration (e.g. coughing), the intrapleural pressure remains negative throughout the cycle because expiration is normally *passive*.

The **dead space** refers to the volume of the airways that does not take part in gas exchange. The **anatomical dead space** includes the respiratory tract down to the terminal bronchioles; it is normally ~150 mL. The **alveolar dead space** refers to the alveoli incapable of gas exchange; in health, it is negligible. The **physiological dead space** is the sum of the anatomical and alveolar dead space.

22 Lung mechanics

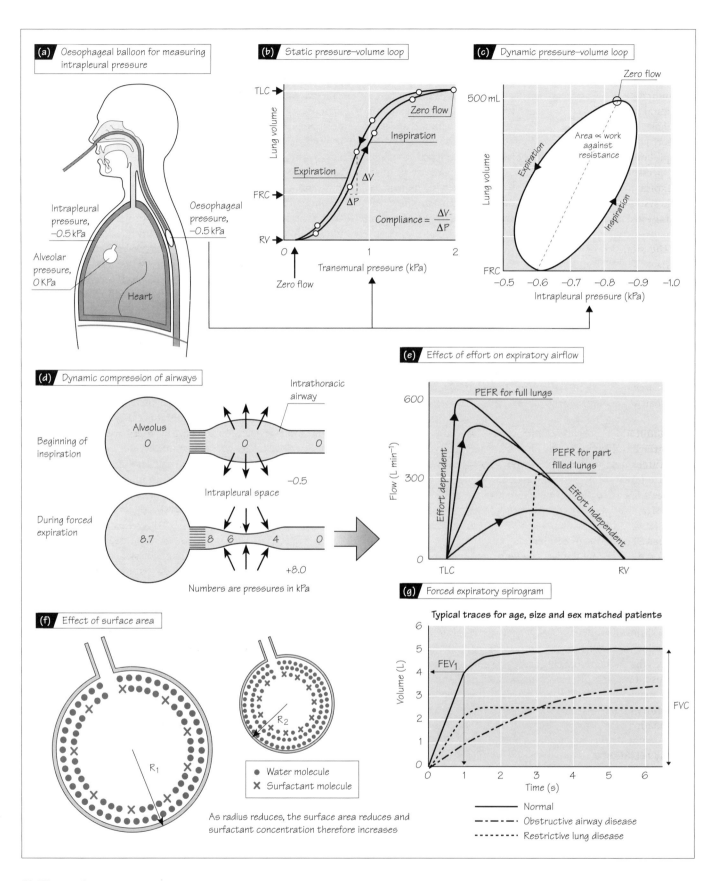

(a) Oesophageal balloon for measuring intrapleural pressure

Intrapleural pressure, −0.5 kPa

Oesophageal pressure, −0.5 kPa

Alveolar pressure, 0 KPa

Heart

(b) Static pressure–volume loop

Lung volume
TLC
FRC
RV

Zero flow
Inspiration
Expiration
ΔV
ΔP

$$Compliance = \frac{\Delta V}{\Delta P}$$

Transmural pressure (kPa)

Zero flow

(c) Dynamic pressure–volume loop

500 mL

Zero flow

Area ∝ work against resistance

Expiration
Inspiration

Lung volume

FRC

Intrapleural pressure (kPa)

(d) Dynamic compression of airways

Intrathoracic airway

Beginning of inspiration

Alveolus 0 — 0 — 0

Intrapleural space −0.5

During forced expiration

8.7 — 8 6 4 0

+8.0

Numbers are pressures in kPa

(e) Effect of effort on expiratory airflow

Flow (L min⁻¹)
600
300
0

PEFR for full lungs
PEFR for part filled lungs
Effort dependent
Effort independent

TLC
RV

(f) Effect of surface area

R₁
R₂

• Water molecule
✕ Surfactant molecule

As radius reduces, the surface area reduces and surfactant concentration therefore increases

(g) Forced expiratory spirogram

Typical traces for age, size and sex matched patients

Volume (L)
6
5
4
3
2
1
0

FEV_1
FVC

Time (s)
0 1 2 3 4 5 6

——— Normal
—·—·— Obstructive airway disease
········ Restrictive lung disease

The respiratory muscles have to overcome resisting forces during breathing. These are primarily the **elastic resistance** in the chest wall and lungs, and the resistance to air flow (**airway resistance**).

Lung compliance

The **static compliance** ('stretchiness') of the lungs (C_L) is defined as the change in volume per unit change in distending pressure ($C_L = \Delta V / \Delta P$) when there is no air flow. The distending pressure is the **transmural** (alveolar–intrapleural) pressure (Chapter 21). The intrapleural pressure can be measured with an oesophageal balloon (Fig. 22a). The alveolar pressure is the same as the mouth pressure (i.e. zero) if no air is flowing. The subject breathes in steps and the intrapleural pressure is measured at each held volume. A typical static **pressure–volume plot** is shown in Fig. 22b. The inspiratory and expiratory curves are slightly different (*hysteresis*), typical for elastic systems. The **static lung compliance** is the maximum slope, generally just above the functional residual capacity (FRC), and is normally ~1.5 L kPa^{-1}, although this is dependent on age, size and sex. The static compliance is reduced by lung *fibrosis* (stiffer lungs).

The **dynamic compliance** is measured during continuous breathing, and therefore includes a component due to **airway resistance**. The dynamic pressure–volume loop (Fig. 22c) has a point at each end where the flow is zero; the slope of the line between these points is the **dynamic compliance**. This is normally similar to the static compliance, but can be altered in disease. The width of the curve reflects the pressure required to suck in or expel air; the *area of the curve* is therefore a measure of the work done against airway resistance.

Surfactant and the alveolar air–fluid interface

The **surface tension** of the fluid lining the alveoli contributes to lung stiffness, as the attraction of water molecules at the air–fluid interface tends to collapse the alveoli. This is a manifestation of **Laplace's law** (Chapter 11), which shows that the pressure in a bubble (or alveolus) is proportional to the surface tension (T) and radius ($P \propto T/r$). A small bubble will therefore have a higher pressure than a larger one and, if connected, will collapse into it. The inward force created by this surface tension also tends to suck fluid into the alveoli (**transudation**). In the lung, these problems are minimized by **surfactant**, secreted by **type II pneumocytes** (Chapter 21). Surfactant is a mixture of **phospholipids** that floats on the alveolar fluid surface, and reduces surface tension. As the alveoli shrink, the effective concentration of surfactant increases, further lowering the surface tension (Fig. 22f). This more than balances the effect of reducing radius (as r falls, so does T). Surfactant also reduces lung stiffness and transudation. Premature babies may not have sufficient surfactant, and develop **neonatal respiratory distress syndrome**, with stiff lungs, lung collapse and transudation.

Airway resistance

Flow through the airways is described by **Darcy's law**, flow = $(P_1 - P_2)/R$ (Chapter 11), where P_1 is the alveolar pressure, P_2 is the mouth pressure and R is the resistance to air flow. The airway resistance is determined by the airway radius, according to **Poiseuille's law**, and whether the flow is laminar or turbulent (Chapter 11).

The airway resistance is increased by factors that constrict the airway smooth muscle (**bronchoconstrictors**). These include the reflex release of **muscarinic** neurotransmitters from parasympathetic nerve endings, generally due to the activation of **irritant receptors** (Chapter 25), and numerous mediators released by inflammatory cells (e.g. **histamine, prostaglandins, leukotrienes**), for example in **asthma**. Increased mucus production also narrows the lumen and increases the resistance. Sympathetic stimulation, adrenaline (epinephrine) and salbutamol cause relaxation and **bronchodilatation** via β_2-adrenoceptors on the smooth muscle.

Effect of transmural pressure. Expiration is normally passive (Chapter 21). Forced expiration increases the intrapleural and thus alveolar pressure, increasing the pressure gradient to the mouth and therefore theoretically leading to increased flow. However, although expiration from fully inflated lungs is indeed **effort dependent**, towards the end of the breath increasing force does not increase flow, i.e. it is **effort independent** (Fig. 22e). This occurs as a result of the pressure gradient between the alveoli and the mouth. Midway between them, generally in the bronchi, the pressure in the airway falls below the intrapleural pressure, causing the airway to collapse (**dynamic compression**; Fig. 22d). As there is now no flow, the pressure rises again until it is greater than the intrapleural pressure, and the airway reopens. This sequence happens repeatedly, producing the brassy sound heard during forced expiration. This does not occur in normal expiration because the intrapleural pressure remains negative throughout. In diseases in which the airways are already narrowed (e.g. *asthma*), this leads to expiratory wheezing and air trapping.

Lung function tests

Lung volumes can be measured using a simple spirometer (Chapter 21). Airway resistance and lung compliance can be assessed indirectly by measuring the forced expiratory flows and volumes. The easiest and quickest measurement is the **peak expiratory flow rate** (PEFR). PEFR is decreased if the airway resistance is increased (**obstructive disease**), and is commonly used to follow an already diagnosed condition, e.g. asthma. It is, however, dependent on the initial lung volume (Fig. 22e). Plots of the **forced expiratory volume against time** provide more information. Subjects breathe out from total lung capacity to residual volume as fast as possible; this is the **forced vital capacity** (FVC), and a typical trace is shown in Fig. 22g. The forced expiratory volume in 1 s (**FEV$_1$**) reflects the airway resistance; it is normally expressed as a ratio to FVC (**FEV$_1$/FVC**) to correct for lung volume, and is usually 0.75–0.90. It can be used to distinguish between **obstructive** (increased airway resistance) and **restrictive** (decreased lung compliance) diseases. In asthma, for example, FEV$_1$/FVC is typically <0.7. In restrictive disease (e.g. lung fibrosis), FEV$_1$ and FVC are low, but FEV$_1$/FVC is normal or even increased due to greater elastic recoil (Fig. 22g).

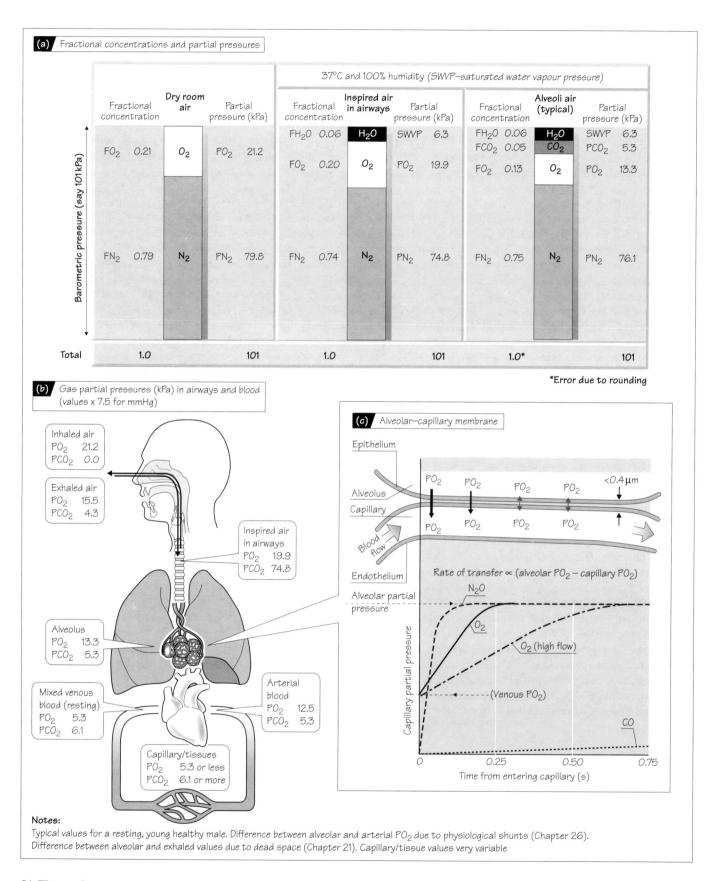

(a) Fractional concentrations and partial pressures

37°C and 100% humidity (SWVP–saturated water vapour pressure)

	Dry room air			Inspired air in airways			Alveoli air (typical)	
Fractional concentration		Partial pressure (kPa)	Fractional concentration		Partial pressure (kPa)	Fractional concentration		Partial pressure (kPa)
			FH_2O 0.06	H_2O	SWVP 6.3	FH_2O 0.06	H_2O	SWVP 6.3
						FCO_2 0.05	CO_2	PCO_2 5.3
FO_2 0.21	O_2	PO_2 21.2	FO_2 0.20	O_2	PO_2 19.9	FO_2 0.13	O_2	PO_2 13.3
FN_2 0.79	N_2	PN_2 79.8	FN_2 0.74	N_2	PN_2 74.8	FN_2 0.75	N_2	PN_2 76.1
Total 1.0		101	1.0		101	1.0*		101

Barometric pressure (say 101 kPa)

*Error due to rounding

(b) Gas partial pressures (kPa) in airways and blood (values x 7.5 for mmHg)

Inhaled air
PO_2 21.2
PCO_2 0.0

Exhaled air
PO_2 15.5
PCO_2 4.3

Inspired air in airways
PO_2 19.9
PCO_2 74.8

Alveolus
PO_2 13.3
PCO_2 5.3

Mixed venous blood (resting)
PO_2 5.3
PCO_2 6.1

Arterial blood
PO_2 12.5
PCO_2 5.3

Capillary/tissues
PO_2 5.3 or less
PCO_2 6.1 or more

(c) Alveolar–capillary membrane

Epithelium

PO_2 PO_2 PO_2 PO_2

<0.4 μm

Alveolus
Capillary

PO_2 PO_2 PO_2 PO_2

Blood flow

Endothelium

Rate of transfer ∝ (alveolar PO_2 – capillary PO_2)

Alveolar partial pressure

N_2O
O_2
O_2 (high flow)
(Venous PO_2)
CO

Capillary partial pressure

0   0.25   0.50   0.75
Time from entering capillary (s)

Notes:
Typical values for a resting, young healthy male. Difference between alveolar and arterial PO_2 due to physiological shunts (Chapter 26).
Difference between alveolar and exhaled values due to dead space (Chapter 21). Capillary/tissue values very variable

Dry air contains 78.1% N_2 and 21% O_2; other inert gases account for the balance (0.9%), but are normally pooled with N_2 (i.e. $N_2 = $ 79%). The small amount of CO_2 in air (<0.04%) is usually ignored.

Partial pressures and fractional concentrations (Fig. 23a)

The volume of a fixed amount of gas is inversely proportional to the pressure ($V \propto 1/P$; **Boyle's law**) and proportional to the absolute temperature ($V \propto T$; **Charles' law**). An ideal gas occupies 22.4 L per mole at 1 atm (101 kPa, 760 mmHg) and 0 °C (273 K), and thus the volume of each gas in a mixture is directly proportional to the quantity of that gas in moles. The term **fractional concentration (F)** can therefore be used to denote the relative quantities of gases in any mixture; thus F_{N_2} is 0.79 in *dry* air, and F_{O_2} is 0.21. The **partial pressure** of each gas in a mixture is that part of the total (e.g. barometric) pressure that is exerted by that gas, and is directly proportional to the quantity. Thus, according to **Dalton's law**, the partial pressure of O_2 (P_{O_2}) in dry air is $F_{O_2} \times$ barometric pressure (P_B), e.g. $0.21 \times 101 \text{ kPa} = 21.2 \text{ kPa}$. At the summit of Everest, P_B is ~34 kPa, but the relative proportions of gases are the same as at sea level, and so P_{O_2} is $0.21 \times 34 \text{ kPa} = 7.14 \text{ kPa}$.

Water vapour pressure. Water vapour behaves like any other gas, and exerts a partial pressure. The maximum or **saturated water vapour pressure** depends on the temperature: 2.33 kPa at 20 °C and 6.3 kPa at 37 °C. Inspired air quickly reaches body temperature and becomes fully humidified (100% saturated) in the airways. Water vapour dilutes the other gases, so that P_{N_2} and P_{O_2} will be lower than in dry air. Thus, P_{O_2} will be $0.21 \times (P_B - $ saturated water vapour pressure) or, under these conditions, $0.21 \times (101 - 6.3) = 23.9 \text{ kPa}$ (Fig. 23a). The water vapour content of room air depends on the conditions (e.g. desert vs. seaside); 40% humidity denotes 40% of the predicted saturated water vapour pressure for that temperature.

Standardization. From the above and Boyle's and Charles' laws, it should be clear that gas volumes and partial pressures cannot be compared unless corrected to a standardized pressure, temperature and humidity. Two standards are commonly used: standard temperature and pressure, dry gas (**STPD**), corrected to 1 standard atm (101 kPa), 0 °C and dry gas; and body temperature and pressure (1 atm), saturated with water (**BTPS**).

Gases dissolved in body fluids

The quantity of gas dissolving in a fluid is described by **Henry's law**: dissolved gas concentration = partial pressure of gas above fluid × **solubility** of that gas in that fluid. The solubility tends to decrease with a rise in temperature, and varies significantly between gases. For example, CO_2 is 20 times more soluble than O_2 in water, so that water exposed to the same partial pressures of CO_2 and O_2 will contain 20 times as much CO_2 as O_2. Henry's law describes an *equilibrium* — increasing the partial pressure of a gas will cause more to dissolve in the fluid until a new equilibrium is reached. The concept of a partial pressure of gas dissolved in a fluid (e.g. P_{O_2} of blood) is sometimes difficult to understand, but merely reflects the partial pressure that would be required to dissolve that amount of gas in the fluid, according to Henry's law. From the above, it can be deduced that the movement of gases between gas and fluid phases (e.g. alveolar air and capillary blood) will be dependent on the **difference in partial pressures** rather than the concentration. Typical values for partial pressures in the airways and blood are shown in Fig. 23b.

Diffusion across the alveolar–capillary membrane (Fig. 23c)

Diffusion is discussed in Chapter 11. The rate of gas flow across the alveolar–capillary membrane = permeability × area × (difference in partial pressures), where the permeability depends on the membrane thickness, gas molecular weight and its solubility in the membrane (Chapter 11). Although CO_2 is larger than O_2, it crosses the membrane faster because it is more soluble in biological membranes. For gas transfer across the lungs, the permeability and area are commonly combined as the **diffusing capacity (D_L)** for that gas, a measure of alveolar–capillary membrane function. Thus, the rate of O_2 transfer $= D_L O_2 \times$ (alveolar P_{O_2} – lung capillary P_{O_2}), or $D_L O_2 = O_2$ **uptake from lungs/(alveolar P_{O_2} – lung capillary P_{O_2})**. $D_L O_2$ is sometimes called the **transfer factor**. $D_L O_2$ cannot be estimated directly, because capillary P_{O_2} cannot be measured. However, the factors affecting O_2 diffusion also affect carbon monoxide (CO) diffusion. CO binds extremely strongly to haemoglobin, and so, if low concentrations of CO are inhaled, CO diffusing into the blood is completely bound to haemoglobin and capillary P_{CO} remains close to zero (Fig. 23c). Thus, $D_L CO = CO$ uptake from lungs/alveolar P_{CO}, and can be easily measured as an estimate of alveolar–capillary transfer function. $D_L CO$ is reduced by a decrease in lung exchange area (e.g. emphysema) or an increase in alveolar–capillary membrane thickness (e.g. lung fibrosis, oedema).

Diffusion and perfusion limitation (Fig. 23c)

Because CO binds so avidly and rapidly to haemoglobin, at low concentrations its rate of transfer into the blood is not affected by the blood flow, because there is always plenty of haemoglobin, but is limited solely by its rate of diffusion across the alveolar–capillary membrane, i.e. transfer is **diffusion limited**. For a poorly soluble gas, however (e.g. the anaesthetic nitrous oxide, N_2O), the partial pressure in the blood rapidly reaches equilibrium with alveolar air, preventing further diffusion. In this case, increased blood flow will increase the rate of transfer, i.e. transfer is **perfusion limited**. O_2 transfer is normally perfusion limited.

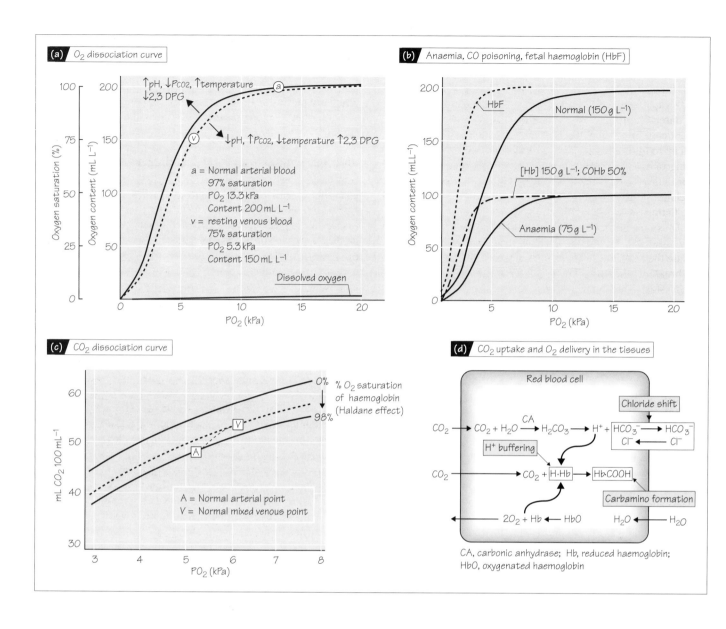

(a) O_2 dissociation curve

↑pH, ↓Pco_2, ↑temperature
↓2,3 DPG

↓pH, ↑Pco_2, ↓temperature ↑2,3 DPG

a = Normal arterial blood
97% saturation
PO_2 13.3 kPa
Content 200 mL L^{-1}
v = resting venous blood
75% saturation
PO_2 5.3 kPa
Content 150 mL L^{-1}

Dissolved oxygen

Oxygen saturation (%)
Oxygen content (mL L^{-1})
PO_2 (kPa)

(b) Anaemia, CO poisoning, fetal haemoglobin (HbF)

HbF
Normal (150 g L^{-1})
[Hb] 150 g L^{-1}; COHb 50%
Anaemia (75 g L^{-1})

Oxygen content (mL L^{-1})
PO_2 (kPa)

(c) CO_2 dissociation curve

0%
% O_2 saturation
of haemoglobin
(Haldane effect)
98%

V
A

A = Normal arterial point
V = Normal mixed venous point

mL CO_2 100 mL^{-1}
PO_2 (kPa)

(d) CO_2 uptake and O_2 delivery in the tissues

Red blood cell

Chloride shift

CO_2 → $CO_2 + H_2O$ →[CA] H_2CO_3 → $H^+ + HCO_3^-$ → HCO_3^-
Cl^- ← Cl^-

H^+ buffering

CO_2 → $CO_2 + H \cdot Hb$ → $Hb \cdot COOH$

Carbamino formation

$2O_2 + Hb$ ← HbO H_2O ← H_2O

CA, carbonic anhydrase; Hb, reduced haemoglobin;
HbO, oxygenated haemoglobin

Oxygen

The resting O_2 consumption in adults is ~250 mL min^{-1}, rising to >4000 mL min^{-1} during heavy exercise. The O_2 **solubility** in plasma is, however, low and at a P_{O_2} of 13 kPa blood contains only 3 mL L^{-1} of dissolved O_2 in solution. Most O_2 is therefore carried bound to **haemoglobin** in red blood cells. Each gram of haemoglobin can combine with 1.34 mL of O_2 and so, for a haemoglobin concentration [Hb] of 150 g L^{-1}, blood can contain a maximum of 200 mL O_2 L^{-1} (**O_2 capacity**). The actual amount of O_2 bound to haemoglobin (**O_2 content**) depends on P_{O_2}, and the **percentage O_2 saturation** = content/capacity×100 (Fig. 24a). Each haemoglobin molecule binds up to four O_2 molecules; binding is **cooperative**, so that the binding of each O_2 molecule makes it easier for the next. This steepens the **O_2 haemoglobin dissociation curve**, which describes the relationship between blood O_2 content and P_{O_2} (Fig. 24a). The curve flattens above ~8 kPa P_{O_2} as all binding sites become occupied. Thus, for a normal arterial P_{O_2} (~13 kPa) and [Hb], the blood is ~97% saturated and contains slightly less than 200 mL O_2 L^{-1}. Because the dissociation curve is flat in this region, any increase in P_{O_2} (breathing O_2-enriched air) will have little effect on content. On the steep part of the curve, however (<8 kPa P_{O_2}), small changes in P_{O_2} will have large effects on content.

Oxygen uptake and delivery. The high P_{O_2} in the lungs facilitates O_2 binding to haemoglobin, whereas the low P_{O_2} in the tissues encourages release. The dissociation curve is shifted to the right (reduced affinity, facilitating O_2 release) by a fall in pH, a rise in P_{CO_2} (**Bohr shift**) and an increase in temperature, which occur in active tissues (Fig. 24a). The metabolic by-product **2,3-diphosphoglycerate** (2,3-DPG) also causes a right shift. In the lungs, P_{CO_2} falls, the pH consequently rises and the temperature is reduced; these all increase affinity and shift the curve to the left, facilitating O_2 uptake.

Anaemia. This is an abnormally low [Hb]; the O_2 capacity is therefore less and the O_2 content at any P_{O_2} is reduced (Fig. 24b). Arterial P_{O_2} and O_2 saturation remain normal. In order to deliver the same amount of O_2 to the tissues, the capillary P_{O_2} would have to fall further than normal (Fig. 24b), reducing the driving force for O_2 diffusion into the tissues. The latter may become inadequate for metabolism, especially during exercise, although a 50% reduction in [Hb] does not usually cause symptoms at rest.

Carbon monoxide. Carbon monoxide (CO) binds 240 times more strongly than O_2 to haemoglobin and, by occupying O_2-binding sites, reduces the O_2 capacity. However, unlike anaemia, CO also increases the affinity and shifts the dissociation curve to the left, making O_2 release to the tissues more difficult. Thus, if 50% of haemoglobin is bound to CO, P_{O_2} needs to fall much further than in anaemia to release the same amount of O_2, causing symptoms of severe hypoxia (headache, convulsions, coma, death) (Fig. 24b).

Fetal haemoglobin. Fetal haemoglobin (HbF) binds 2,3-DPG less strongly than does adult haemoglobin (HbA), and so the dissociation curve is shifted to the left. This facilitates the transfer of O_2 from maternal blood to the fetus, where the arterial P_{O_2} is only ~5 kPa (Fig. 24b).

Carbon dioxide

Blood can carry much more CO_2 than O_2, and CO_2 is transported as **bicarbonate, carbamino compounds** and simply **dissolved** in plasma. The **CO_2 dissociation** curve does not plateau and is more linear than the O_2 dissociation curve (Fig. 24c).

Bicarbonate. Approximately 60% of CO_2 is carried as bicarbonate. Water and CO_2 combine to form carbonic acid (H_2CO_3) and thence bicarbonate (HCO_3^-): $CO_2 + H_2O \Leftrightarrow H_2CO_3 \Leftrightarrow HCO_3^- + H^+$. The left side of the equation is normally slow, but speeds up dramatically in the presence of **carbonic anhydrase**, found in red cells. Bicarbonate is therefore formed preferentially in red cells, from which it easily diffuses out. Red cells are, however, impermeable to H^+ ions, and Cl^- enters the cell to maintain electrical neutrality (**chloride shift**) (Fig. 24d). H^+ binds avidly to **deoxygenated** (*reduced*) haemoglobin (haemoglobin acts as a buffer), and so there is little increase in [H^+] to impede further bicarbonate formation. *Oxygenated* haemoglobin does not bind H^+ as well, and so in the lungs H^+ dissociates from haemoglobin and shifts the CO_2–HCO_3^- equation to the left, assisting CO_2 unloading from the blood (Fig. 24d); the reverse occurs in the tissues. This contributes to the **Haldane effect**, which states that, for any P_{CO_2}, the CO_2 content of oxygenated blood is less than that of deoxygenated blood.

Carbamino compounds. These compounds are formed by the reaction of CO_2 with protein amino groups: $CO_2 + \text{protein } NH_2 \Leftrightarrow \text{protein NH COOH}$. The most prevalent protein in blood is haemoglobin, which forms **carbaminohaemoglobin** with CO_2. This occurs more readily for deoxygenated than oxygenated haemoglobin, contributing to the Haldane effect (Fig. 24d). Carbamino compounds account for 30% of CO_2 carriage.

Dissolved carbon dioxide. CO_2 is 20 times more soluble than O_2 in plasma, and ~10% of CO_2 in blood is carried in **solution**.

Hyperventilation and hypoventilation

Doubling the rate of ventilation halves the alveolar and arterial P_{CO_2}. Ventilation is normally closely matched to the metabolic rate as reflected by CO_2 production (Chapter 25). **Hyperventilation** (overventilation) and **hypoventilation** (underventilation) are defined in terms of arterial P_{CO_2}, so that a subject is *hyperventilating* when $P_{CO_2} < 5.3$ kPa, and hypoventilating when $P_{CO_2} > 5.9$ kPa. Rapid breathing in exercise is *not* hyperventilation, as this is appropriate for increased CO_2 production and P_{CO_2} does not fall. Hyperventilation cannot normally increase the O_2 content, as arterial haemoglobin is already nearly fully saturated. The fall in P_{CO_2} (**hypocapnia**) during hyperventilation causes light-headedness, visual disturbances due to cerebral vasoconstriction (Chapter 20) and muscle cramps (tetany). Hyperventilation can be caused by pain, hysteria and strong emotion. Hypoventilation causes a high P_{CO_2} (**hypercapnia**) and a low P_{O_2} (**hypoxia**), and may be caused by head injury or respiratory disease.

25 Control of breathing

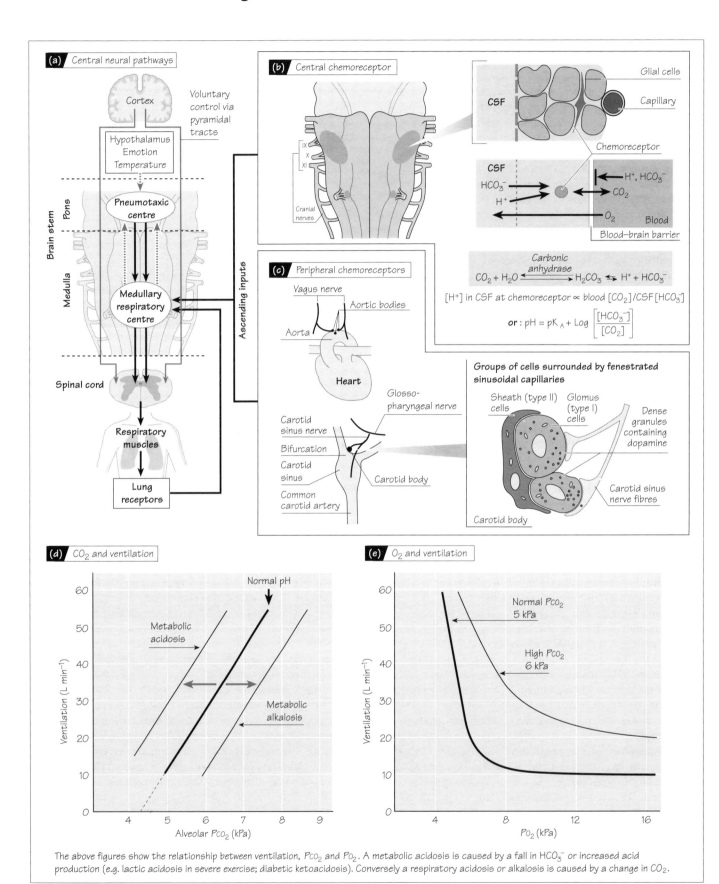

(a) Central neural pathways

Cortex

Voluntary control via pyramidal tracts

Hypothalamus
Emotion
Temperature

Brain stem

Pons

Pneumotaxic centre

Medulla

Medullary respiratory centre

Ascending inputs

Spinal cord

Respiratory muscles

Lung receptors

(b) Central chemoreceptor

ix
x
xi

Cranial nerves

Glial cells

Capillary

CSF

Chemoreceptor

CSF

HCO_3^-
H^+

H^+, HCO_3^-
CO_2
O_2

Blood

Blood–brain barrier

Carbonic anhydrase

$$CO_2 + H_2O \rightleftharpoons H_2CO_3 \rightleftharpoons H^+ + HCO_3^-$$

$[H^+]$ in CSF at chemoreceptor $\propto$ blood $[CO_2]$/CSF $[HCO_3^-]$

or : $pH = pK_A + \text{Log} \left[\dfrac{[HCO_3^-]}{[CO_2]} \right]$

(c) Peripheral chemoreceptors

Vagus nerve

Aortic bodies

Aorta

Heart

Glosso-pharyngeal nerve

Carotid sinus nerve

Bifurcation

Carotid sinus

Common carotid artery

Carotid body

Groups of cells surrounded by fenestrated sinusoidal capillaries

Sheath (type II) cells

Glomus (type I) cells

Dense granules containing dopamine

Carotid sinus nerve fibres

Carotid body

(d) CO_2 and ventilation

Normal pH

Metabolic acidosis

Metabolic alkalosis

Ventilation (L min^{-1})

Alveolar P_{CO_2} (kPa)

(e) O_2 and ventilation

Normal P_{CO_2} 5 kPa

High P_{CO_2} 6 kPa

Ventilation (L min^{-1})

P_{O_2} (kPa)

The above figures show the relationship between ventilation, P_{CO_2} and P_{O_2}. A metabolic acidosis is caused by a fall in HCO_3^- or increased acid production (e.g. lactic acidosis in severe exercise; diabetic ketoacidosis). Conversely a respiratory acidosis or alkalosis is caused by a change in CO_2.

Ventilation of the lungs provides O_2 for the tissues and removes CO_2. Breathing must therefore be closely matched to metabolism for adequate O_2 delivery and to prevent a build-up of CO_2. Other factors must also be addressed, such as requirements for speech and coughing, and the current state of the lungs and respiratory muscles. All of these are integrated by the **respiratory centre** located in the brain stem. This sets the basic rhythm and pattern of ventilation, and is modulated by **chemoreceptors** to match ventilation to metabolism, feedback from **lung receptors** and higher inputs, such as emotion and temperature.

Respiratory centre

The respiratory centre is not a single entity, but encompasses diffuse groups of neurones in the **pons** and **medulla** (Fig. 25a). The **medullary respiratory centre** contains the **rhythm generator** for normal breathing, and is modulated by ascending input from **chemoreceptors** and **lung receptors**, and descending input via the **pneumotaxic centre** in the pons. The **pneumotaxic centre** receives input from the hypothalamus and higher centres, and also from the medullary centre itself. It coordinates the medullary homeostatic functions with factors such as emotion and temperature, and affects the pattern of breathing. Voluntary control is mediated by cortical motor neurones in the **pyramidal tract**, which bypasses the respiratory centres.

Chemoreception

Chemoreceptors detect arterial P_{CO_2}, P_{O_2} and pH — P_{CO_2} is most important. Alveolar P_{CO_2} ($P_{A}CO_2$) is normally ~5.3 kPa (40 mmHg), and $P_{A}O_2$ normally 13 kPa (100 mmHg). An increase in $P_{A}CO_2$ causes ventilation to rise in an almost linear fashion (Fig. 25d). Increased acidity of the blood (e.g. lactic acidosis in severe exercise) causes the relationship between P_{CO_2} and ventilation to shift to the right, and decreased acidity causes a shift to the left. Conversely, P_{O_2} normally only stimulates ventilation when it falls below ~8 kPa (~60 mmHg) (Fig. 25e). However, when a fall in P_{O_2} is accompanied by an increase in P_{CO_2}, the resultant increase in ventilation is greater than would be expected from the effects of either alone; there is thus a **synergistic** (more than additive) relationship between P_{O_2} and P_{CO_2} (Fig. 25e).

The **central chemoreceptor** comprises a collection of neurones near the ventrolateral surface of the medulla, close to the exit of the IXth and Xth cranial nerves (Fig. 25b). It responds *indirectly* to blood P_{CO_2}, but does **not** respond to changes in P_{O_2}. Although CO_2 can easily diffuse across the **blood–brain barrier** from the blood into the cerebrospinal fluid (CSF), H^+ and HCO_3^- cannot. As a result, the pH of the CSF around the chemoreceptor is determined by the arterial P_{CO_2} and CSF $[HCO_3^-]$, according to the Henderson–Hasselbalch equation (Fig. 25b). A rise in blood P_{CO_2} therefore makes the CSF more acid; this is detected by the chemoreceptor, which increases ventilation to blow off CO_2. The central chemoreceptor is responsible for ~80% of the response to CO_2 in humans. Its response is delayed because CO_2 has to diffuse across the blood–brain barrier. As the blood–brain barrier is impermeable to H^+, the central chemoreceptor is not affected by blood pH.

The **peripheral chemoreceptors** are located in the carotid and aortic bodies (Fig. 25c). The **carotid bodies** are small distinct structures located at the bifurcation of the common carotid arteries, and are innervated by the carotid sinus nerve and thence the glossopharyngeal. The carotid body is formed from **glomus** (type I) and **sheath** (type II) cells. Glomus cells are chemoreceptive, contain dopamine-rich dense granules and contact carotid sinus nerve axons. The **aortic bodies** are located on the aortic arch and are innervated by the vagus. They are similar to carotid bodies but functionally less important. Peripheral chemoreceptors respond to changes in P_{CO_2}, $[H^+]$ and, importantly, P_{O_2}. They are responsible for ~20% of the response to increased P_{CO_2}.

Lung receptors

Various types of lung receptor provide feedback from the lungs to the respiratory centre. In addition, **pain** often causes brief apnoea (cessation of breathing) followed by rapid breathing, and **mechanical** or **noxious** stimulation of receptors in the trigeminal region and larynx causes apnoea or spasm of the larynx.

Stretch receptors. These are located in the bronchial walls. Stimulation (by stretch) causes short, shallow breaths, and delay of the next inspiratory cycle. They provide negative feedback to turn off inspiration. They are mostly **slowly adapting** (continue to fire with sustained stimulation) and are innervated by the vagus. They are largely responsible for the **Hering–Breuer inspiratory reflex**, in which lung inflation inhibits inspiration to prevent overinflation.

Juxtapulmonary (J) receptors. These are located on the alveolar and bronchial walls close to the capillaries. They cause depression of somatic and visceral activity by producing rapid shallow breathing or apnoea, a fall in heart rate and blood pressure, laryngeal constriction and relaxation of the skeletal muscles via spinal neurones. They are stimulated by increased alveolar wall fluid, oedema, microembolisms and inflammation. The afferent nerves are small unmyelinated (C-fibre) or myelinated nerves in the vagus.

Irritant receptors. These are located throughout the airways between epithelial cells. In the trachea they cause cough, and in the lower airways hyperpnoea (rapid breathing); stimulation also causes bronchial and laryngeal constriction. They are also responsible for the deep augmented breaths every 5–20 min at rest, reversing the slow collapse of the lungs that occurs in quiet breathing, and may be involved in the first deep gasps of the newborn. They are stimulated by irritant gases, smoke and dust, rapid large inflations and deflations, airway deformation, pulmonary congestion and inflammation. The afferent nerves are rapidly adapting myelinated fibres in the vagus.

Proprioceptors (position/length sensors). These are located in the Golgi tendon organs, muscle spindles and joints. They are important for matching increased load, and maintaining optimal tidal volume and frequency. They are stimulated by shortening and load in the respiratory muscles (but not the diaphragm). Afferents run to the spinal cord via the dorsal roots. It should be noted that input from non-respiratory muscles and joints can also stimulate breathing.

26 Ventilation–perfusion matching and right to left shunts

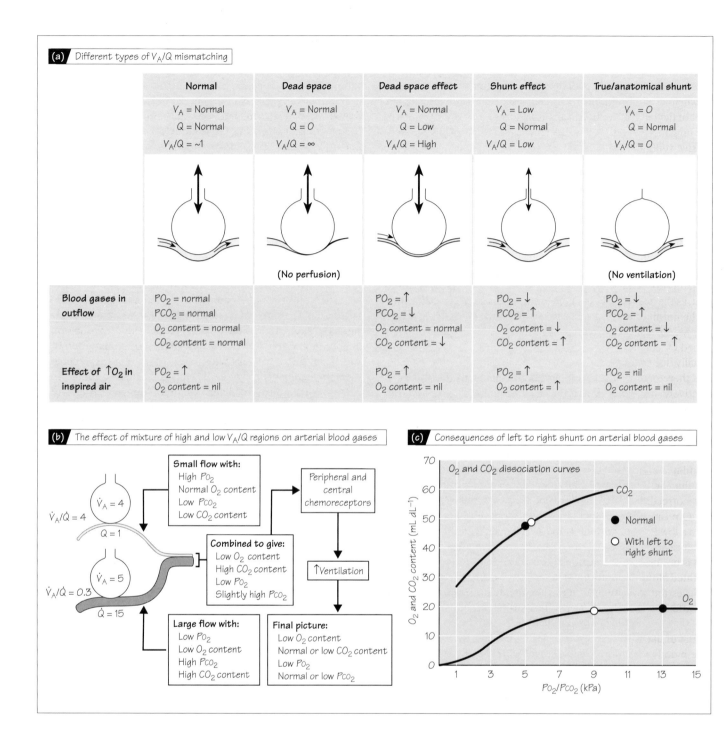

(a) Different types of V_A/Q mismatching

	Normal	Dead space	Dead space effect	Shunt effect	True/anatomical shunt
	V_A = Normal	V_A = Normal	V_A = Normal	V_A = Low	V_A = 0
	Q = Normal	Q = 0	Q = Low	Q = Normal	Q = Normal
	V_A/Q = ~1	V_A/Q = ∞	V_A/Q = High	V_A/Q = Low	V_A/Q = 0
		(No perfusion)			(No ventilation)
Blood gases in outflow	PO_2 = normal PCO_2 = normal O_2 content = normal CO_2 content = normal		PO_2 = ↑ PCO_2 = ↓ O_2 content = normal CO_2 content = ↓	PO_2 = ↓ PCO_2 = ↑ O_2 content = ↓ CO_2 content = ↑	PO_2 = ↓ PCO_2 = ↑ O_2 content = ↓ CO_2 content = ↑
Effect of ↑O_2 in inspired air	PO_2 = ↑ O_2 content = nil		PO_2 = ↑ O_2 content = nil	PO_2 = ↑ O_2 content = ↑	PO_2 = nil O_2 content = nil

(b) The effect of mixture of high and low V_A/Q regions on arterial blood gases

$\dot{V}_A/\dot{Q}$ = 4
$\dot{V}_A$ = 4
Q = 1

$\dot{V}_A/\dot{Q}$ = 0.3
$\dot{V}_A$ = 5
$\dot{Q}$ = 15

Small flow with:
High PO_2
Normal O_2 content
Low PCO_2
Low CO_2 content

Large flow with:
Low PO_2
Low O_2 content
High PCO_2
High CO_2 content

Combined to give:
Low O_2 content
High CO_2 content
Low PO_2
Slightly high PCO_2

Peripheral and central chemoreceptors

↑Ventilation

Final picture:
Low O_2 content
Normal or low CO_2 content
Low PO_2
Normal or low PCO_2

(c) Consequences of left to right shunt on arterial blood gases

O_2 and CO_2 dissociation curves

CO_2

O_2

● Normal
○ With left to right shunt

O_2 and CO_2 content (mL dL^{-1})

PO_2/PCO_2 (kPa)

Ventilation–perfusion matching (Fig. 26a)

At rest, total alveolar ventilation (V_A) is similar to total pulmonary capillary perfusion (Q), or about $5\,L\,min^{-1}$. For optimal gas exchange, all regions of the lung should ideally have a **ventilation–perfusion ratio** (V_A/Q) of unity. When there are variations significantly away from unity, either lower or higher, this is referred to as **ventilation–perfusion mismatch**. In a **right to left shunt** (see below), for example, ventilation is zero and $V_A/Q = \infty$; whereas when an embolism blocks a pulmonary artery, perfusion in the part of the lung fed by that artery is zero, and $V_A/Q = 0$. Regions of the lung that have a V_A/Q value much greater than unity have excessive ventilation, and blood derived from them will have a high Po_2 and a low Pco_2 (**dead space effect**). Regions with a V_A/Q value much less than unity have a **shunt** or **venous admixture** effect; there is some gas exchange, but the blood has a lower than normal Po_2 and a higher than normal Pco_2.

Effect of ventilation–perfusion mismatch on arterial gases. Regions of high V_A/Q cannot compensate for regions of low V_A/Q. This is because of the way in which O_2 is carried in the blood (Chapter 24). Although regions of high V_A/Q produce blood with a high Po_2, this does not translate to any significant increase in O_2 content, as the haemoglobin in the blood is already close to saturation at the normal Po_2. Conversely, blood derived from regions with a low V_A/Q, especially if $Po_2 < 8\,kPa$, will have a significantly reduced O_2 content (Fig. 26c). Regions of high V_A/Q are also most usually due to insufficient perfusion, so that the amount of blood, and therefore O_2, that such regions contribute to the total will be relatively small. As a result, the combined blood from regions with high and low V_A/Q will have a low O_2 content and a low Po_2, even if total ventilation and perfusion are matched for the whole lung.

The CO_2 content is less severely affected, because overventilated areas can lose extra CO_2 and partly compensate for underventilated areas. Moreover, a rise in Pco_2 will stimulate breathing via the chemoreceptors, allowing CO_2 to be corrected, or even overcorrected if Po_2 is sufficiently low (Chapter 25). Significant V_A/Q mismatching will therefore usually result in arterial blood with a low Po_2 but normal or low Pco_2 (Fig. 26b). Ventilation with O_2-enriched air will improve oxygenation in regions of low V_A/Q, but is not useful for shunts, as the enriched air never reaches the shunted blood. **Hypoxic pulmonary vasoconstriction** (Chapter 20) reduces the severity of V_A/Q mismatch by diverting blood from the affected region to well-ventilated areas.

Effect of gravity

The blood pressure at the base of the lungs is greater than that at the apex (top) because of the weight of the column of blood. This increased pressure distends the pulmonary blood vessels at the base and the flow is therefore increased. Conversely, blood flow at the apex may be reduced if the pulmonary venous pressure falls below the alveolar pressure, when the vessels will be compressed. The net result is that, on standing, pulmonary blood flow falls progressively on moving from the bottom of the lung to the top.

Gravity also affects the intrapleural pressure, which is thus less negative at the base of the lung than at the apex. Alveoli at the base are therefore less expanded at functional residual capacity, and thus have more potential for expansion during inspiration. As a result, ventilation is greatest at the base of the lung. Although the effects of gravity on perfusion and ventilation partly cancel each other out, ventilation is less affected than perfusion, so that V_A/Q is highest at the apex of the lung and lowest at the base. In the young, this relatively small variation has little effect on blood gases, but in the elderly, it may contribute to a low Po_2.

Right to left shunts

Part of the venous effluent of the bronchial and coronary circulations bypasses the lungs and enters the pulmonary vein and left ventricle, respectively (Chapter 12). Oxygenated blood from the lungs is therefore diluted by venous blood. These are **anatomical right to left shunts**, and account for <2% of cardiac output in healthy individuals. Larger shunts can occur in disease when regions of the lung are not ventilated (e.g. lung collapse, pneumonia), or due to **congenital heart malformations**. When calculating the effects of right to left shunts on arterial blood, the blood **content** of O_2 and CO_2 needs to be considered. For a 20% shunt, the 80% of blood passing through the lungs will have normal arterial O_2 and CO_2 contents of 200 and $480\,mL\,L^{-1}$, respectively, whilst the 20% bypassing the lungs will have normal venous values of 150 and $520\,mL\,L^{-1}$, respectively. On combination, the blood will contain $(200 \times 0.8) + (150 \times 0.2) = 190\,mL\,L^{-1}$ O_2, and $(480 \times 0.8) + (520 \times 0.2) = 488\,mL\,L^{-1}$ CO_2. From the dissociation curves (Fig. 26c), it can be seen that this results in a fall in Po_2 from 13 to 9 kPa, whereas Pco_2 rises only marginally from 5.3 to 5.5 kPa because of the steeper curve. Changes in Pco_2 and Po_2 stimulate the chemoreceptors and increase ventilation (Chapter 25), so that arterial Pco_2 returns to normal. However, increased ventilation cannot increase blood O_2 content, as the haemoglobin of the blood passing through the lungs is already close to saturation. Thus, right to left shunts commonly result in a *low arterial Po_2* but a *normal or low Pco_2*.

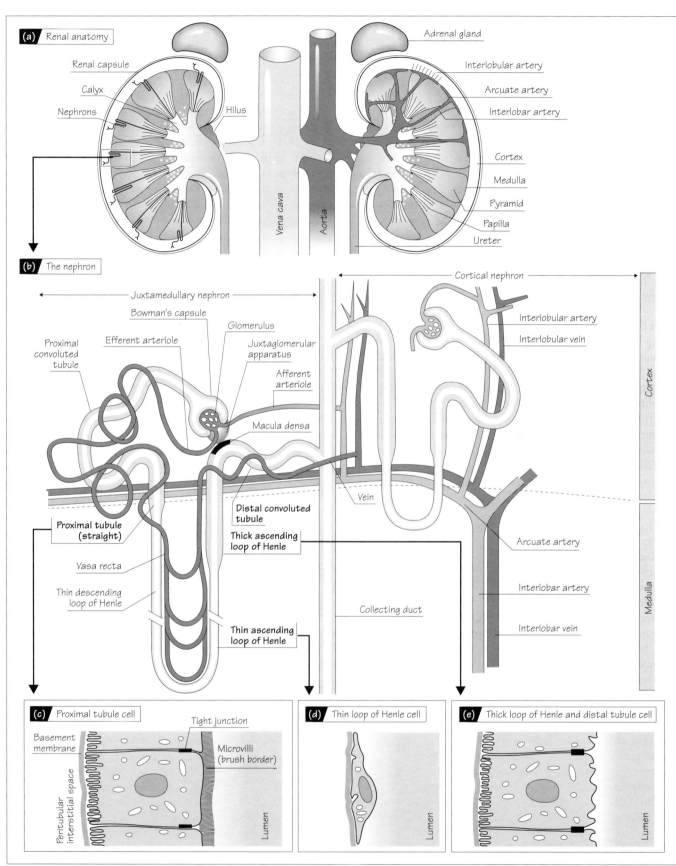

(a) Renal anatomy

Adrenal gland
Renal capsule
Interlobular artery
Calyx
Arcuate artery
Nephrons
Interlobar artery
Hilus
Cortex
Medulla
Pyramid
Papilla
Ureter
Vena cava
Aorta

(b) The nephron

Cortical nephron
Juxtamedullary nephron
Bowman's capsule
Glomerulus
Proximal convoluted tubule
Efferent arteriole
Juxtaglomerular apparatus
Interlobular artery
Interlobular vein
Afferent arteriole
Macula densa
Distal convoluted tubule
Proximal tubule (straight)
Thick ascending loop of Henle
Vein
Vasa recta
Arcuate artery
Thin descending loop of Henle
Collecting duct
Interlobar artery
Thin ascending loop of Henle
Interlobar vein
Cortex
Medulla

(c) Proximal tubule cell
Tight junction
Basement membrane
Microvilli (brush border)
Peritubular interstitial space
Lumen

(d) Thin loop of Henle cell
Lumen

(e) Thick loop of Henle and distal tubule cell
Lumen

The kidneys help to maintain the composition of extracellular body fluids, and regulate ions (e.g. Na^+, K^+, Ca^{2+}, Mg^{2+}), acid–base status and body water. They also have an endocrine function. Plasma is **filtered** by capillaries in the **glomerulus** (Chapter 28), and the composition of the filtrate is modified by **reabsorption** and **secretion** in the nephrons. The average urine output is ~1.5 L per day, although this can fall to <1 L per day and increase to nearly 20 L per day.

Gross structure

The kidneys are located on each side of the vertebral column, behind the peritoneum. The renal artery and vein, lymphatics and nerve enter the kidney via the **hilus**, from which the **renal pelvis**, which becomes the **ureter**, emerges (Fig. 27a). The kidney is surrounded by a fibrous **renal capsule**. Internally, the kidney has a dark outer **cortex** surrounding a lighter **medulla**, which contains triangular lobes or **pyramids**. The cortex contains the **glomerulus** and **proximal** and **distal tubules** of the **nephrons**, whilst the **loop of Henle** and **collecting ducts** descend into the medulla (Fig. 27b). Each kidney contains ~800 000 nephrons. The collecting ducts converge in the **papilla** at the apex of each pyramid, and empty into the **calyx** (plural: *calyces*) and thence renal pelvis. Urine is propelled through the ureter into the **bladder** by peristalsis.

The nephron

Each **nephron** begins with a capsule (**Bowman's capsule**) surrounding the glomerular capillaries, which collects filtrate (Fig. 28a), followed by the **proximal tubule**, **loop of Henle**, **distal tubule** and early **collecting duct** (Fig. 27b). There are two types of nephron—those with glomeruli in the outer 70% of the cortex and short loops of Henle (**cortical nephrons**: ~85%), and those with glomeruli close to the cortex–medulla boundary and long loops of Henle (**juxtamedullary nephrons**: ~15%).

The **glomerulus** produces **ultrafiltrate** from plasma (Chapter 28).

The **proximal tubule** is convoluted when it leaves Bowman's capsule, but straightens before becoming the descending limb of the loop of Henle in the medulla. Its walls are formed from *columnar epithelial* cells with a **brush-border** of *microvilli* on the lumenal surface that increases the surface area ~40-fold (Fig. 27c). **Tight junctions** close to the lumenal side limit diffusion through gaps between cells. The basal or peritubular side of the cells shows considerable *interdigitation*, which increases the surface area. The term **lateral intercellular space** is often used to describe the space between the interdigitations and basement membrane and between the base of adjacent cells. The main function of the proximal tubule is **reabsorption** (Chapter 29).

The thin part of the **loop of Henle** (~20 μm across) is formed from thin, flat (*squamous*) cells (Fig. 27d), with no microvilli. The **thick ascending loop of Henle** has columnar epithelial cells similar to the proximal tubule, but with few microvilli (Fig. 27e). At the point at which the loop associates with the **juxtaglomerular apparatus** (see Chapter 31), after re-entering the cortex, the wall is formed from modified **macula densa** cells (Fig. 27b). The loop of Henle is important for the production of concentrated urine.

The **distal tubule** is functionally similar to the **cortical collecting duct**. Both contain cells similar to those in the thick ascending loop of Henle (Fig. 27e). In the collecting duct, these **principal cells** are interspersed with **intercalated cells** of different morphology and function; these play a role in acid–base balance (see Chapter 32). The collecting duct plays an important role in water homeostasis (Chapter 31).

Renal circulation

The kidneys receive ~20% of cardiac output. The renal artery enters via the hilus and divides into **interlobar arteries** running between the pyramids to the cortex–medulla boundary, where they split into **arcuate arteries**. **Interlobular arteries** ascend into the cortex, and feed the **afferent arterioles** of the glomerulus (Fig. 27a,b). The capillaries of the glomerulus are the site of **filtration**, and drain into the **efferent arteriole** (*not* vein). Afferent and efferent arterioles provide the major resistance to renal blood flow. Efferent arterioles branch into a network of capillaries in the cortex around the proximal and distal tubules (**peritubular capillaries**). Capillaries close to the cortex–medulla boundary loop into the medulla to form the **vasa recta** surrounding the loop of Henle; this provides the only blood supply to the medulla. All capillaries drain into the renal veins. Ninety per cent of the blood entering the kidney supplies the cortex, giving a high blood flow (~500 mL $min^{-1} (100 g)^{-1}$) and a low arteriovenous O_2 difference (~2%). Medullary blood flow is less (20–100 mL $min^{-1} (100 g)^{-1}$).

Regulation of renal blood flow. Differential constriction of afferent and efferent arterioles strongly affects filtration (see above; Chapter 28). The kidneys exhibit a high degree of **autoregulation** (Fig. 28e), both by the **myogenic** response (Chapter 20) and via the macula densa, which detects high filtration rates and releases adenosine, which constricts afferent arterioles, so reducing filtration. Noradrenaline (norepinephrine) from renal sympathetic nerves constricts both afferent and efferent arterioles, and increases renin and thus the production of angiotensin II (a potent vasoconstrictor) (Chapter 31). Many peripheral vasoconstrictors (e.g. endothelin, angiotensin II) cause the release of vasodilating prostaglandins in the kidney, so protecting renal blood flow.

Hormones and the kidney

Renal function is affected by a variety of hormones that modulate the regulation of ions and water (e.g. **antidiuretic hormone**, **aldosterone**). **Renin** is produced by the juxtaglomerular apparatus and promotes the formation of angiotensin (Chapter 31). **Erythropoietin** is synthesized by interstitial cells in the cortex, and stimulates red cell production (Chapter 9). **Vitamin D** is metabolized in the kidney to its active form (**1,25-dihydroxycholecalciferol**), which is involved in Ca^{2+} and phosphate regulation (Chapters 30 & 44). Various **prostaglandins** are also produced in the kidney, and affect renal blood flow.

Micturition

The constriction of smooth muscle in the bladder wall (**detrusor** muscle) expels urine through the **urethra** (**micturition**, urination). Micturition is initiated by a spinal reflex when urine pressure reaches a critical level, but is strongly controlled by higher (voluntary) centres. The neck of the bladder forms the **internal urethral sphincter**; the **external sphincter** is formed from voluntary skeletal muscle around more distal regions of the urethra.

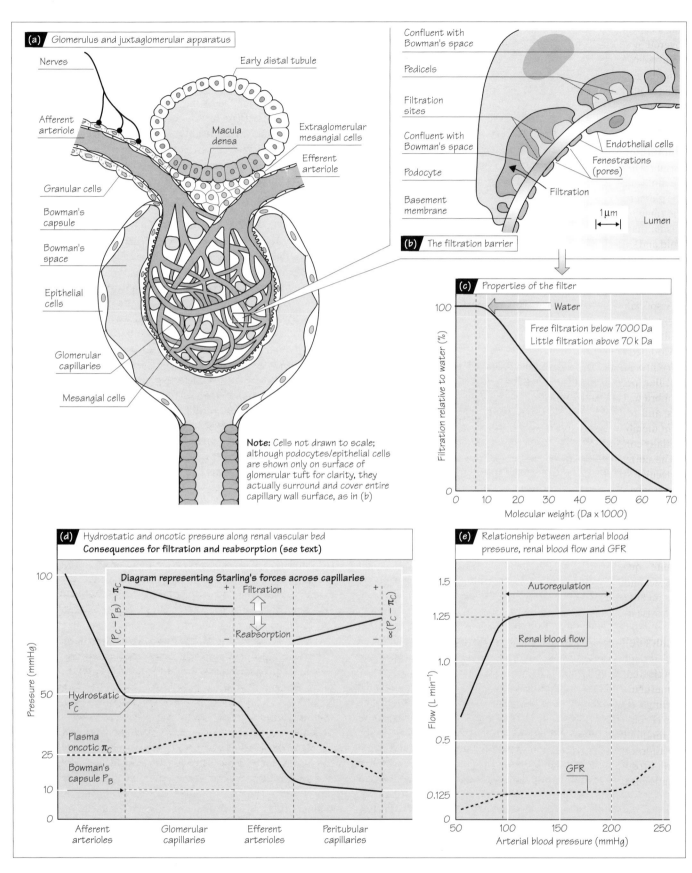

(a) Glomerulus and juxtaglomerular apparatus

Nerves
Early distal tubule
Afferent arteriole
Macula densa
Extraglomerular mesangial cells
Efferent arteriole
Granular cells
Bowman's capsule
Bowman's space
Epithelial cells
Glomerular capillaries
Mesangial cells

Note: Cells not drawn to scale; although podocytes/epithelial cells are shown only on surface of glomerular tuft for clarity, they actually surround and cover entire capillary wall surface, as in (b)

Confluent with Bowman's space
Pedicels
Filtration sites
Confluent with Bowman's space
Podocyte
Basement membrane
Endothelial cells
Fenestrations (pores)
Filtration
1 µm
Lumen

(b) The filtration barrier

(c) Properties of the filter

Water
Free filtration below 7000 Da
Little filtration above 70 k Da

Filtration relative to water (%)
Molecular weight (Da x 1000)

(d) Hydrostatic and oncotic pressure along renal vascular bed
Consequences for filtration and reabsorption (see text)

Diagram representing Starling's forces across capillaries

$(P_C - P_B) - \pi_C$
Filtration
Reabsorption
$\propto (P_C - \pi_C)$

Pressure (mmHg)

Hydrostatic P_C

Plasma oncotic π_C

Bowman's capsule P_B

Afferent arterioles
Glomerular capillaries
Efferent arterioles
Peritubular capillaries

(e) Relationship between arterial blood pressure, renal blood flow and GFR

Autoregulation

Renal blood flow

GFR

Flow (L min^{-1})
Arterial blood pressure (mmHg)

The structure of the **glomerulus** is shown in Fig. 28a. The walls of the afferent arteriole are associated with *granular cells* that produce **renin** (Chapter 31); there are numerous sympathetic nerve endings. The tuft of glomerular capillaries is surrounded by **Bowman's capsule**, the inner surface of which and the capillaries are covered by specialized epithelial cells (**podocytes**; see below). The glomerulus is interspersed with **mesangial** cells which are **phagocytic** (engulf large molecules) and **contractile**; contraction may limit the filtration area and alter filtration. Mesangial cells are also found between the capsule and **macula densa** (*extraglomerular mesangial cells*; Fig. 28a).

Glomerular filtration

Plasma is filtered in the glomerulus by **ultrafiltration** (i.e. works at the molecular level), and filtrate passes into the proximal tubule. The **glomerular filtration rate** (GFR) is $\sim125\,mL\,min^{-1}$ in humans. The renal *plasma* flow is $\sim600\,mL\,min^{-1}$, so that the proportion of plasma that filters into the nephron (**filtration fraction**) is $\sim20\%$. Fluid and solutes have to pass three filtration barriers (Fig. 28b):

1 The **glomerular capillary endothelium**, which is approximately 50 times more permeable than in most tissues because it is **fenestrated** with small (70 nm) pores (Chapter 19).

2 A specialized capillary **basement membrane** containing negatively charged glycoproteins, which is thought to be the main site of ultrafiltration.

3 Modified epithelial cells (**podocytes**) with long extensions (*primary processes*) that engulf the capillaries and have numerous foot-like processes (**pedicels**) directly contacting the basement membrane. The regular gaps between pedicles are called **filtration slits**, and restrict large molecules. Podocytes maintain the basement membrane and, like mesangial cells, may be phagocytic and partially contractile.

The permeability of the filtration barrier is dependent on the molecular size. Substances with molecular weights of <7000 Da pass freely, but larger molecules are increasingly restricted up to 70 000–100 000 Da, above which filtration is insignificant (Fig. 28c). Negatively charged molecules are further restricted as they are repelled by negative charges in the basement membrane. Thus, albumin ($\sim69\,000$ Da), which is also negatively charged, is filtered in minute quantities, whereas small molecules such as ions, glucose, amino acids and urea pass the filter without hindrance. This means that the glomerular filtrate is almost protein free, but otherwise has an identical composition to plasma.

Factors determining the glomerular filtration rate

GFR is dependent on the difference between the **hydrostatic** and **oncotic** (colloidal osmotic, due to proteins) pressures in the glomerular capillaries and Bowman's capsule, as determined by **Starling's equation** (Chapter 19). The glomerular capillary pressure (P_c) is greater than that elsewhere (~48 mmHg) because of the unique arrangement of afferent and efferent arterioles and low afferent but high efferent resistances. As the pressure in Bowman's capsule (P_B) is ~10 mmHg, the net hydrostatic force driving filtration is (P_c-P_B) or ~35 mmHg. This is opposed by the oncotic

pressure of capillary plasma (π_c; ~25 mmHg); the filtrate oncotic pressure is essentially zero (no protein). Thus, $GFR \propto (P_c - P_B) - \pi_c$ (Fig. 28d). It should be noted that, because the filtration fraction is appreciable ($\sim20\%$) and proteins are not filtered, the plasma protein concentration and thus π_c will rise as blood traverses the glomerulus, reducing (*but not abolishing*) filtration. In peritubular capillaries, where the hydrostatic pressure is very low, this increase in π_c promotes reabsorption (Fig. 28d).

GFR is therefore strongly dependent on the relative resistance of afferent and efferent arterioles, which is influenced by sympathetic tone and other vasoactive agents. GFR is constant over a wide range of blood pressure (90–200 mmHg) because of the **autoregulation** of renal blood flow (Fig. 28e; Chapter 20). Renal disease, circulating and local vasoconstrictors, and sympathetic activation all reduce GFR, although angiotensin II preferentially constricts *efferent* arterioles, and thus increases GFR (see Chapter 31).

Measurement of the glomerular filtration rate and the concept of clearance

If substance X is freely filtered and neither reabsorbed nor secreted in the nephron, the amount appearing in the urine per minute must equal the amount filtered per minute. Thus, if the plasma concentration of X is C_p and the urine concentration is C_u, and the volume of urine passed per minute is V, then $C_p \times GFR = C_u \times V$, or **$GFR = (C_u \times V)/C_p$**. **Creatinine**, which is steadily released from skeletal muscle, is often used for clinical measurements of GFR because it is freely filtered and not reabsorbed; there is a little secretion, but this introduces only a small error, except when plasma creatinine or GFR is abnormally low. More accurate measurements are made by infusing the polysaccharide **inulin**, which is neither reabsorbed nor secreted.

This is known as a **clearance method**. The term **clearance** can be confusing, as it does not refer to what actually happens but is merely a way of looking at how the kidney deals with a substance. It is defined as the volume of plasma that would need to be completely cleared of a substance per minute in order to produce the amount found in the urine, or: **clearance $= (C_u \times V)/C_p$** (i.e. the same equation as above). Thus, the **clearance of inulin is equal to GFR**. If a substance is reabsorbed in the nephron, its clearance will be less than GFR and, if it is secreted, it will be greater than GFR. Some substances that are normally completely reabsorbed have zero clearance until the reabsorption mechanism becomes saturated (e.g. glucose; Chapter 29).

The **renal plasma flow** (RPF) can be measured in a similar fashion by infusing *para*-**aminohippuric acid (PAH)** which at low concentrations is completely removed from renal blood by both filtration and secretion, so that none remains in the venous outflow. The amount appearing in the urine must therefore equal the amount entering the kidney, and thus the **clearance of PAH is equal to RPF**. The filtration fraction (GFR/RPF; see above) can therefore be estimated from inulin clearance/PAH clearance. The renal blood flow is equal to RPF/(1 – haematocrit).

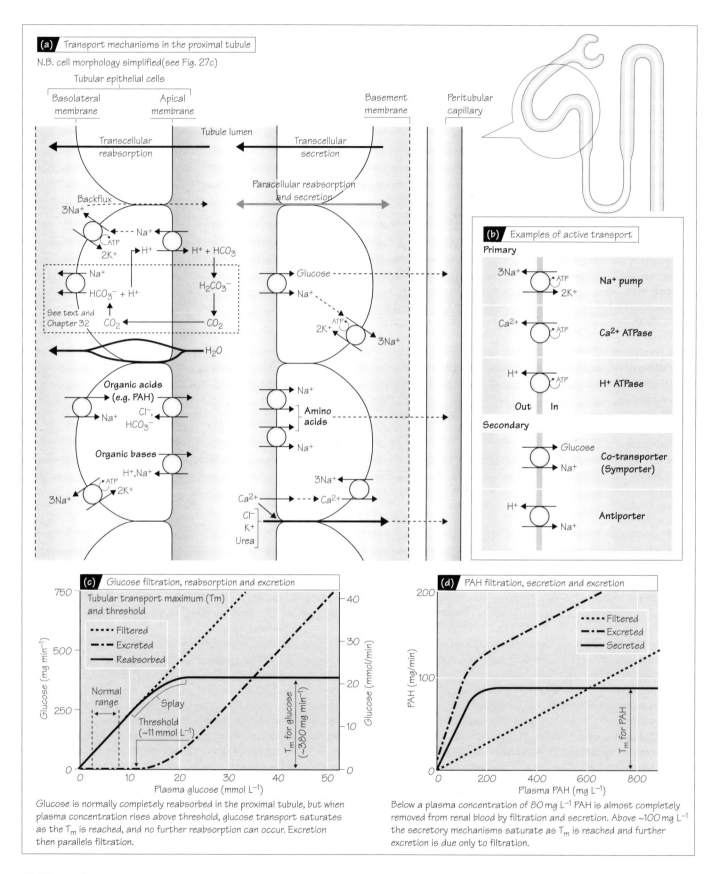

(a) Transport mechanisms in the proximal tubule

N.B. cell morphology simplified(see Fig. 27c)

(b) Examples of active transport

(c) Glucose filtration, reabsorption and excretion

Glucose is normally completely reabsorbed in the proximal tubule, but when plasma concentration rises above threshold, glucose transport saturates as the T_m is reached, and no further reabsorption can occur. Excretion then parallels filtration.

(d) PAH filtration, secretion and excretion

Below a plasma concentration of 80 mg L^{-1} PAH is almost completely removed from renal blood by filtration and secretion. Above ~100 mg L^{-1} the secretory mechanisms saturate as T_m is reached and further excretion is due only to filtration.

In a healthy adult, ~180 L of filtrate enters the proximal tubules daily. A significant component must be reabsorbed to prevent the loss of water and solutes. The filtrate is progressively modified as it passes through the nephron by the **reabsorption** of substances into the blood and **secretion** into the tubular fluid. The net reabsorption or secretion of any substance can be determined from its **clearance** (Chapter 28).

Tubular transport processes

Reabsorption and secretion involve the transport of substances across the tubular epithelium; this occurs either by diffusion through tight junctions and lateral intercellular spaces (**paracellular** pathway), driven by concentration, osmotic or electrical gradients, or by **active transport** through the epithelial cells themselves (**transcellular** pathways) (Fig. 29a). The latter usually involves an active process on either the apical or basolateral cell membrane, with passive diffusion across the opposite membrane driven by the concentration gradient so created. The movement of solutes between the peritubular space and capillaries is by **bulk flow and diffusion** (Chapter 11); the movement of water is influenced by **Starling's forces** (Chapter 19).

Active transport involves proteins called **transporters** that translocate substances across the cell membrane (Fig. 29b; Chapter 5). **Primary active transport** uses adenosine triphosphate (ATP) directly, e.g. the **Na^+-K^+-ATPase** (**Na^+ pump**). **Secondary active transport** uses the concentration gradient created by primary active transport as an energy source. This is most commonly the Na^+ gradient created by the *Na^+ pump*, and the latter therefore plays a critical role in renal reabsorption and secretion. **Symporters** (or *cotransporters*) transport substances in the same direction as (for example) Na^+, whereas **antiporters** transport in the opposite direction (Chapter 5; Fig. 29b).

The rate of *diffusion* across cell membranes is enhanced by **ion channels** and **uniporters** (transporters carrying only one substance), which effectively increase membrane permeability to specific substances; this is termed **facilitated diffusion**, and may be modulated by hormones or drugs.

Tubular transport maximum

There is a limit to the rate at which any transporter can operate, and so, for any substance, there is a maximum rate of reabsorption or secretion, called the **tubular transport maximum (T_m)**. For example, glucose is normally completely reabsorbed in the proximal tubule and none is excreted in the urine (see below). However, when the filtrate glucose concentration rises above the **renal threshold**, the transporters start to **saturate**, and glucose appears in the urine (Fig. 29c). Once T_m is reached, excretion increases linearly with filtration. The threshold concentration is somewhat lower than that required to reach T_m because of the variation in transport maxima between nephrons; this is called **splay**. Secretory mechanisms also exhibit T_m. For example, at low concentrations, *para*-aminohippuric acid (PAH) is almost completely removed from capillary blood by filtration and secretion (Chapter 28). At higher concentrations secretion becomes saturated, and further excretion is limited to the filtered load (Fig. 29d).

The proximal tubule (Fig. 29a)

The majority of glucose, amino acids, phosphate and bicarbonate is reabsorbed in the proximal tubule, together with 60–70% of Na^+, K^+, Ca^{2+}, urea and water. The secretion of H^+ and reabsorption of HCO_3^- are discussed in detail in Chapter 32.

Sodium. The concentration of Na^+ in the filtrate is ~140 mmol L^{-1}, but is much lower in the cytosol of epithelial cells (~10–20 mmol L^{-1}), which is also negatively charged. The electrochemical gradient therefore favours the movement of Na^+ from the filtrate into the cells, providing the driving force for the secondary transport of other substances. About 80% of Na^+ entering proximal tubular cells exchanges for H^+ (Na^+–H^+ antiporter). The secretion of H^+ in the proximal tubule plays a critical role in HCO_3^- reabsorption (Fig. 29a; Chapter 32). Na^+ is removed from tubular cells by Na^+ pumps primarily on the basolateral membrane, thus transporting Na^+ into the interstitial fluid. However, only ~20% of transported Na^+ diffuses into the capillaries, as there is significant backflux into the tubule via paracellular pathways.

Water. Water is not actively reabsorbed. As Na^+ and HCO_3^- are transported from the tubule into the peritubular interstitial fluid, the **osmolality** of the latter increases, whilst that of the tubular fluid decreases. This osmotic pressure difference causes the reabsorption of water via both transcellular and paracellular pathways.

The reabsorption of water increases tubular concentrations of Cl^-, K^+, Ca^{2+} and **urea**, which therefore diffuse down their concentration gradients into the peritubular space, largely via paracellular pathways, although the route for Ca^{2+} may be transcellular. The final two-thirds of the proximal tubule has increased permeability to Cl^-, facilitating Cl^- reabsorption. This makes the lumen more positive, enhancing the reabsorption of cations. As the reabsorption of Na^+, Cl^-, K^+, Ca^{2+} and urea in the proximal tubule is closely coupled to the reabsorption of water, their concentrations (and the total osmolality) are similar in the fluid leaving the proximal tubule to those in the filtrate and plasma, although their total quantity and fluid volume are decreased by ~70%.

Glucose. Glucose is reabsorbed by **cotransport** with Na^+ across the apical membrane of epithelial cells, and then diffuses out of the cells into the peritubular interstitium. The T_m for glucose is ~380 mg min^{-1} (~21 mmol min^{-1}), and the renal threshold is ~11 mmol L^{-1}. The appearance of glucose in the urine reflects **hyperglycaemia** (high plasma glucose), a sign of **diabetes mellitus**.

Amino acids. Amino acids are reabsorbed by several Na^+-linked symporters, specific for acidic, basic and neutral amino acids.

Phosphate. Phosphate is cotransported with Na^+ across the epithelial apical membrane. Its T_m is close to the filtered load, and so an increase in plasma concentration leads to excretion. Phosphate reabsorption is decreased by **parathyroid hormone**.

Organic acids and bases. These include metabolites (e.g. bile salts, urate, oxalate) and drugs (e.g. PAH, penicillins, aspirin), and are secreted. Organic acids are transported from the peritubular fluid into tubular cells by cotransport with Na^+, and diffuse into the tubule in exchange for anions (e.g. Cl^-, HCO_3^-). Organic bases are actively extruded from the apical membrane in exchange for Na^+ or H^+.

30 The loop of Henle and distal nephron

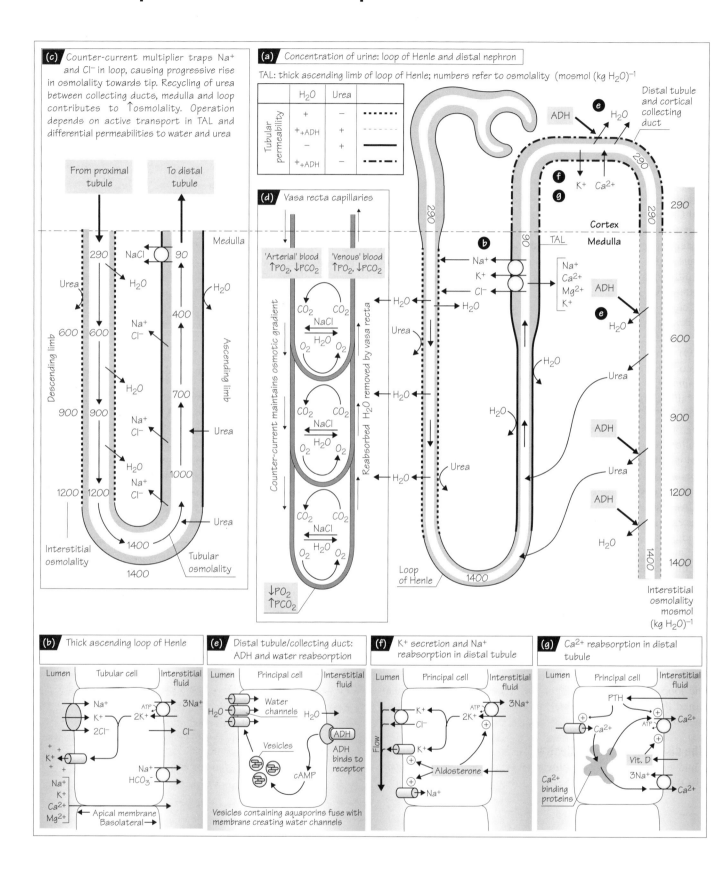

(c) Counter-current multiplier traps Na⁺ and Cl⁻ in loop, causing progressive rise in osmolality towards tip. Recycling of urea between collecting ducts, medulla and loop contributes to ↑osmolality. Operation depends on active transport in TAL and differential permeabilities to water and urea.

(a) Concentration of urine: loop of Henle and distal nephron

TAL: thick ascending limb of loop of Henle; numbers refer to osmolality (mosmol (kg H_2O)⁻¹

	H_2O	Urea	
Tubular permeability	+	−	········
	++ADH	+	-------
	−	+	─────
	++ADH	−	─·─·─

(d) Vasa recta capillaries

(b) Thick ascending loop of Henle

(e) Distal tubule/collecting duct: ADH and water reabsorption

Vesicles containing aquaporins fuse with membrane creating water channels

(f) K⁺ secretion and Na⁺ reabsorption in distal tubule

(g) Ca²⁺ reabsorption in distal tubule

The loop of Henle and distal nephron allow urine to be **concentrated** through the creation of a **high osmolality** in the medulla, which drives the reabsorption of water from the **collecting ducts**. The distal nephron also regulates K^+ and Ca^{2+} excretion and acid–base status (Chapter 32).

The loop of Henle

Fluid entering the descending limb of the loop of Henle is isotonic with plasma (~290 mosmol $(kg H_2O)^{-1}$). The generation of high osmolality in the medulla depends on the **differential permeabilities** to water and solutes in different regions, the **active transport** of ions in the thick ascending limb and the **counter-current multiplier**. The **thin descending limb** is permeable to water but impermeable to urea, whereas the **ascending limb** is impermeable to water but permeable to urea (Fig. 30a); it is also very highly permeable to Na^+ and Cl^-. The **thick ascending limb** actively reabsorbs Na^+ and Cl^- from the tubular fluid by means of apical **Na^+–K^+–$2Cl^-$ cotransporters**; Na^+ is primarily transported across the basolateral membrane by Na^+ pumps (some by Na^+–HCO_3^- cotransport), and Cl^- by diffusion (Fig. 30b). K^+ leaks back into the lumen via apical K^+ channels, creating a positive charge that drives the reabsorption of cations (**$Na^+, K^+, Ca^{2+}, Mg^{2+}$**) through paracellular pathways. As the thick ascending limb is impermeable to water, the reabsorption of ions reduces the tubular fluid osmolality (to ~90 mosmol $(kg H_2O)^{-1}$) and increases the interstitial fluid osmolality, creating an osmotic difference of **~200 mosmol $(kg H_2O)^{-1}$**.

Counter-current multiplier (Fig. 30c). The increased interstitial osmolality causes water to diffuse out of the descending limb, and some Na^+ and Cl^- to diffuse in, concentrating the tubular fluid (Fig. 30c). As this concentrated fluid descends, it travels in the opposite direction to fluid returning from the still higher osmolality regions of the deep medulla. This **counter-current** arrangement creates an osmotic gradient, causing Na^+ and Cl^- to diffuse out of the ascending limb (diluting the ascending fluid), and water to diffuse out of the descending limb (further concentrating the descending fluid). This effect is potentiated by the fact that the ascending limb is impermeable to water, but highly permeable to Na^+ and Cl^-, and also by the recycling of **urea** between the collecting ducts and ascending limb, which makes an important contribution to urine concentration (see below). At the tip of the loop of Henle, the interstitial fluid can reach an osmolality of **~1400 mosmol $(kg H_2O)^{-1}$**, due in equal parts to NaCl and urea.

The blood supply to the medulla is prevented from dissipating the osmotic gradient between the cortex and medulla by the *counter-current exchanger* arrangement of the **vasa recta** capillaries (Fig. 30d). The vasa recta also removes water reabsorbed from the loop of Henle and medullary collecting ducts. It should be noted that O_2 and CO_2 are also conserved, so that, in the deep medulla, P_{O_2} is low and P_{CO_2} is high.

The distal tubule and collecting duct

Fluid entering the distal tubule is *hypotonic* (~90 mosmol $(kg H_2O)^{-1}$). The distal tubule and cortical collecting duct are impermeable to urea. They are also impermeable to water, except in the presence of **antidiuretic hormone** (**ADH**, *vasopressin*) (Chapter 31), which causes water channels (**aquaporins**) to insert into the apical membrane (Fig. 30e). In the presence of ADH, water diffuses into the cortical interstitium, and the tubular fluid becomes concentrated, reaching a maximum osmolality of ~290 mosmol $(kg H_2O)^{-1}$ (i.e. isotonic with plasma). However, the fluid differs from plasma as large quantities of Na^+, K^+, Cl^- and HCO_3^- have been reabsorbed, their place having being taken by **urea**. This is concentrated as water is reabsorbed, because the distal tubule and cortical collecting duct are impermeable to urea.

The **medullary collecting duct** also becomes permeable to water in the presence of ADH. Water is reabsorbed due to the high osmolality of the medullary interstitium (Fig. 30a). The final urine osmolality can therefore reach **1400 mosmol $(kg H_2O)^{-1}$** under conditions of maximum ADH stimulation; in the absence of ADH, urine is *dilute* (**~60 mosmol $(kg H_2O)^{-1}$**). Although only 15% of nephrons have loops of Henle that pass deep into the medulla, and so contribute to the high medullary osmolality (Chapter 27), the *collecting ducts of all nephrons pass through the medulla and therefore concentrate urine.*

Urea. The **medullary collecting duct** is permeable to urea, which diffuses down its concentration gradient into the medulla and then into the ascending loop of Henle (Fig. 30a). Urea is therefore 'trapped' and partially recycled, so maintaining a high concentration and providing ~50% of the osmolality in the medulla (see above). ADH increases the permeability of the medullary collecting duct to urea and hence its reabsorption by activating epithelial **uniporters** (*facilitated diffusion*); this further increases the medullary osmolality and allows the production of more concentrated urine.

Potassium. Potassium has largely been reabsorbed by the time the distal tubule is reached, and so excretion is regulated by secretion in the late distal tubule. K^+ is actively transported into principal cells by basolateral Na^+ pumps, and passively secreted via **channels** and **K^+–Cl^- cotransport** (Fig. 30f). Secretion is therefore driven by the concentration gradient between the cytosol and tubular fluid. However secreted K^+ will reduce the gradient unless it is washed away, and so **K^+ excretion is increased as tubular flow increases**. Diuretics therefore often lead to K^+ loss (Chapter 32). K^+ secretion is increased by **aldosterone**, which enhances Na^+ pump activity and apical membrane K^+ permeability (Chapter 31). Perturbations of K^+ homeostasis are often associated with acid–base disorders (Chapter 32).

Calcium. Calcium reabsorption in the distal tubule is regulated by **parathyroid hormone** (**PTH**) and **1,25-dihydroxycholecalciferol** (active form of **vitamin D**). PTH activates Ca^{2+} entry channels in the epithelial apical membrane, and a basolateral Ca^{2+}-ATPase that is also activated by 1,25-dihydroxycholecalciferol. Ca^{2+} removal is assisted by an Na^+–Ca^{2+} antiporter. Ca^{2+}-binding proteins prevent cytosolic free Ca^{2+} from rising detrimentally (Fig. 30g). PTH also inhibits phosphate reabsorption (Chapter 29). Ca^{2+} regulation is discussed in Chapter 44.

31 Regulation of plasma osmolality and fluid volume

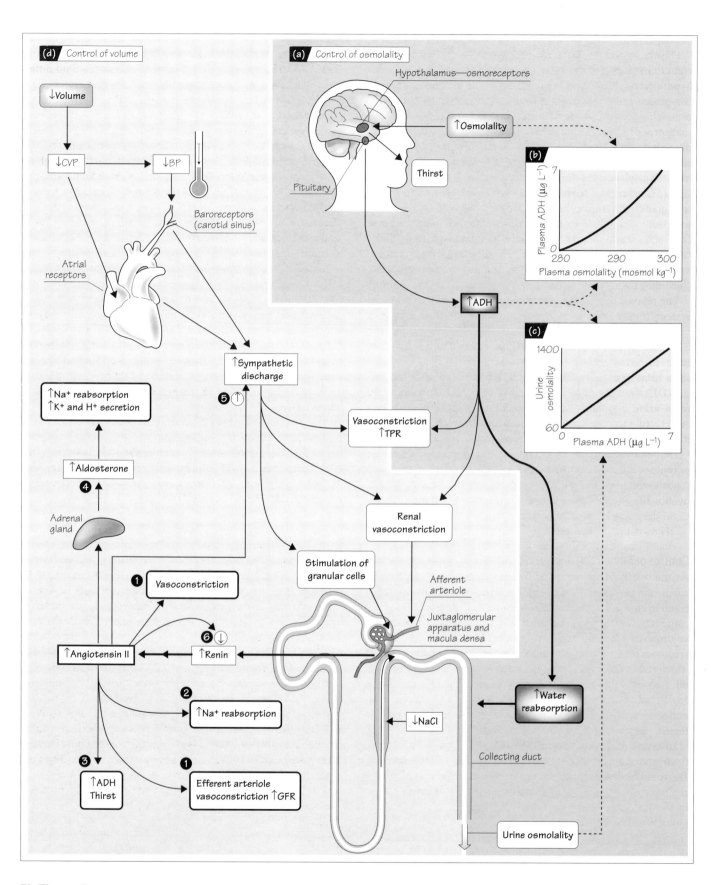

Control of plasma osmolality (Fig. 31a)

Extracellular fluid osmolality must be closely regulated, as alterations cause the swelling or shrinking of all cells, and can lead to cell death. The control of osmolality takes precedence over the control of body fluid volume.

Plasma osmolality is increased in water deficiency and decreased by the ingestion of water. **Osmoreceptors** in the **anterior hypothalamus** are sensitive to changes as small as 1% of plasma osmolality, and regulate **antidiuretic hormone (ADH)**. A rise in osmolality increases ADH release and stimulates **thirst** and water reabsorption (Chapter 30); a fall has the opposite effect. ADH is a nine-amino-acid peptide formed from a large precursor synthesized in the **hypothalamus** (Chapter 40). ADH is transported from here to the **posterior pituitary** (neurohypophysis) within nerve fibres (hypothalamohypophyseal tract), where it is stored in **secretory granules**. Action potentials from osmoreceptors cause these to release ADH. ADH binds to V_2 receptors on renal principal cells and increases cyclic adenosine monophosphate (cAMP), causing the incorporation of water channels (aquaporins) into the apical membrane (Chapter 30). ADH also causes **vasoconstriction** (including renal) via V_1 receptors.

The relationship between plasma osmolality and ADH release is steep (Fig. 31b), as is that between plasma ADH and urine osmolality (Fig. 31c). Normal urine production is ~60 mL h^{-1} (urine osmolality, ~300–800 mosmol (kg H$_2$O)$^{-1}$). Maximum ADH reduces the urine volume to a **minimum** of ~400 mL per day (**maximum urine osmolality**, ~1400 mosmol (kg H$_2$O)$^{-1}$); in the absence of ADH, the urine volume can reach ~25 L per day with a **minimum urine osmolality** of ~60 mosmol (kg H$_2$O)$^{-1}$ (Chapter 30). ADH is rapidly removed from plasma, falling by ~50% in ~10 min, mainly due to metabolism in the liver and kidneys.

Diabetes insipidus is the production of copious amounts of **hypotonic** (dilute) urine due to defective ADH-dependent water reabsorption. This may be due to a congenital defect in ADH production (central diabetes insipidus, CDI), or to a failure to respond to ADH (nephrogenic diabetes insipidus, NDI) due to defective ADH receptors or aquaporins.

Control of body fluid volume (Fig. 31d)

As the osmolality is regulated by the osmoreceptors and ADH, changes in the major component of extracellular fluid, i.e. Na$^+$, will result in changes in extracellular volume. The control of body Na$^+$ content by the kidney is therefore the main regulator of body fluid volume. Atrial and other low-pressure (cardiopulmonary) stretch receptors (Fig. 31d) detect a fall in central venous pressure (**CVP**), which reflects the blood volume. A fall in volume sufficient to reduce blood pressure activates the **baroreceptor reflex** (Chapter 18). In both cases, increased sympathetic discharge causes peripheral vasoconstriction, including vasoconstriction of the **renal afferent arterioles**, stimulation of **ADH release** and water reabsorption (see above), and the **release of renin** (see below) from **granular cells** in the juxtaglomerular apparatus (Chapter 27). **Decreased pressure** in the renal afferent arterioles also stimulates renin release, as does reduced NaCl delivery to the **macula densa** in the juxtaglomerular apparatus (Chapter 27) and a reduced glomerular filtration rate (GFR).

Renin, angiotensin and aldosterone

Renin cleaves plasma angiotensinogen into angiotensin I, which is converted by **angiotensin-converting enzyme (ACE)** on endothelial cells (primarily in the lung) into **angiotensin II**. Angiotensin II is the primary hormone for Na$^+$ homeostasis, and has several important functions (Fig. 31d). ❶ It is a potent **vasoconstrictor** throughout the vasculature, although in the kidney it preferentially constricts efferent arterioles, thereby increasing GFR (Chapter 28) and protecting GFR from a fall in perfusion pressure. ❷ It directly increases **Na$^+$ reabsorption** in the proximal tubule by stimulating Na$^+$–H$^+$ antiporters (Chapter 29). ❸ It stimulates the hypothalamus to increase **ADH secretion** and also causes **thirst**. ❹ It stimulates the production of **aldosterone** by the adrenal cortex. Angiotensin II also tends to ❺ potentiate sympathetic activity (positive feedback) and ❻ inhibit renin production by granular cells (negative feedback). **ACE inhibitors** are important for the treatment of heart failure, when the response to reduced blood pressure leads to detrimental fluid retention and oedema (Chapter 19).

Aldosterone is required for normal Na$^+$ reabsorption and K$^+$ secretion. It increases the synthesis of transport mechanisms in the distal nephron, including the Na$^+$ pump, Na$^+$–H$^+$ symporter and K$^+$ and Na$^+$ channels in principal cells, and H$^+$-ATPase in intercalated cells. Na$^+$ reabsorption and K$^+$ and H$^+$ secretion are thereby enhanced (Chapters 30 & 32). As aldosterone acts via **protein synthesis**, it takes hours to have any effect. The production of aldosterone by the adrenal cortex is directly sensitive to small changes in **plasma [K$^+$]**, suggesting a primary role for K$^+$ homeostasis.

Atrial natriuretic peptide (ANP; atrial natriuretic factor) is released from atrial muscle cells in response to stretch caused by increased blood volume. ANP suppresses the production of renin, aldosterone and ADH, inhibits the effects of ADH in the distal nephron, and causes renal vasodilatation. The net result is the increased excretion of water and Na$^+$.

Diuretics

Osmotic diuretics (e.g. mannitol) cannot be reabsorbed effectively and, consequently, their concentration in tubular fluid increases as water is reabsorbed, limiting further water reabsorption. In **diabetes mellitus**, high plasma glucose saturates glucose reabsorption (Chapter 29), resulting in copious amounts of isotonic urine (i.e. same osmolality as plasma) containing glucose. **Diuretic drugs** generally inhibit tubular transport mechanisms. The most potent are **loop diuretics** (e.g. furosemide), which inhibit Na$^+$–K$^+$–2Cl$^-$ symporters in the thick ascending loop of Henle, thus preventing the development of high osmolality in the medulla and inhibiting water reabsorption (Chapter 30). The increased flow (and thus increased K$^+$ secretion), coupled with reduced K$^+$ reabsorption, enhances K$^+$ excretion and can cause **hypokalaemia** (low plasma [K$^+$]). **Aldosterone antagonists** (e.g. spironolactone) and **Na$^+$ channel blockers** (e.g. amiloride) reduce Na$^+$ entry in the distal nephron and inhibit K$^+$ and H$^+$ secretion; they are weak diuretics, but **K$^+$ sparing**, and are often given with loop diuretics to reduce K$^+$ loss. **Alcohol** inhibits ADH release, and so promotes diuresis.

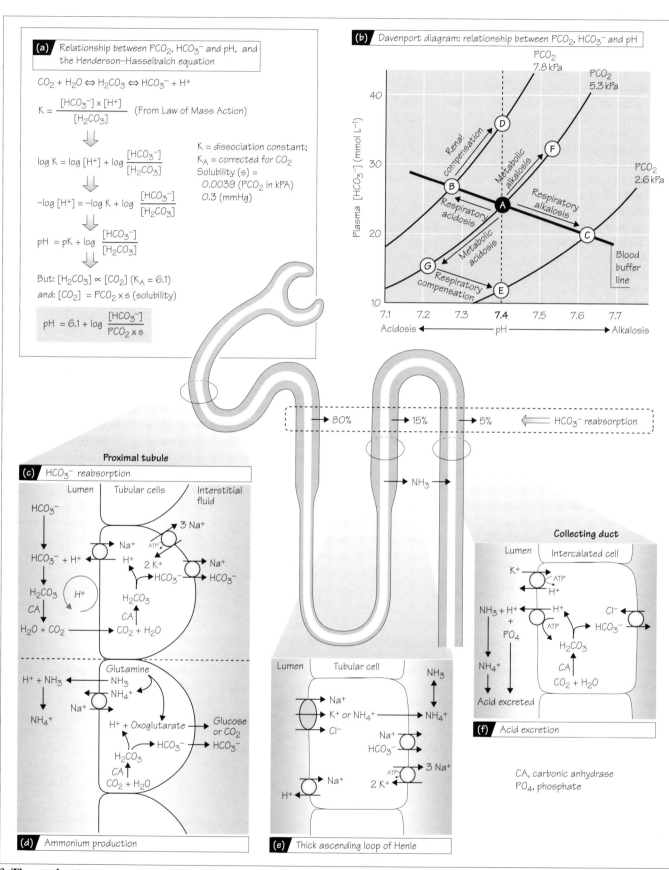

(a) Relationship between PCO_2, HCO_3^- and pH, and the Henderson–Hasselbalch equation

$$CO_2 + H_2O \Leftrightarrow H_2CO_3 \Leftrightarrow HCO_3^- + H^+$$

$$K = \frac{[HCO_3^-] \times [H^+]}{[H_2CO_3]} \quad \text{(From Law of Mass Action)}$$

$$\log K = \log [H^+] + \log \frac{[HCO_3^-]}{[H_2CO_3]}$$

$$-\log [H^+] = -\log K + \log \frac{[HCO_3^-]}{[H_2CO_3]}$$

$$pH = pK + \log \frac{[HCO_3^-]}{[H_2CO_3]}$$

K = dissociation constant;
K_A = corrected for CO_2
Solubility (s) =
   0.0039 (PCO_2 in kPA)
   0.3 (mmHg)

But: $[H_2CO_3] \propto [CO_2]$ ($K_A = 6.1$)
and: $[CO_2] = PCO_2 \times s$ (solubility)

$$pH = 6.1 + \log \frac{[HCO_3^-]}{PCO_2 \times s}$$

(b) Davenport diagram: relationship between PCO_2, HCO_3^- and pH

Plasma $[HCO_3^-]$ (mmol L^{-1})

PCO_2 7.8 kPa
PCO_2 5.3 kPa
PCO_2 2.6 kPa

Renal compensation
Metabolic alkalosis
Respiratory acidosis
Respiratory alkalosis
Metabolic acidosis
Respiratory compensation
Blood buffer line

Acidosis ← pH → Alkalosis

HCO$_3^-$ reabsorption → 80% → 15% → 5% ⇐ HCO_3^- reabsorption

Proximal tubule

(c) HCO_3^- reabsorption

Lumen | Tubular cells | Interstitial fluid

HCO_3^-
$HCO_3^- + H^+$
H_2CO_3
CA
$H_2O + CO_2$

Na^+
H^+
H_2CO_3
CA
$CO_2 + H_2O$

3 Na^+
ATP
2 K^+
HCO_3^-
Na^+
HCO_3^-

NH_3 →

(d) Ammonium production

$H^+ + NH_3$
Na^+
NH_4^+

Glutamine
NH_3
NH_4^+
$H^+ + $ Oxoglutarate
H_2CO_3
CA
$CO_2 + H_2O$
HCO_3^-
Glucose or CO_2
HCO_3^-

(e) Thick ascending loop of Henle

Lumen | Tubular cell

Na^+
K^+ or NH_4^+
Cl^-

NH_3
NH_4^+

Na^+
HCO_3^-
ATP
3 Na^+
2 K^+

Na^+
H^+

Collecting duct

(f) Acid excretion

Lumen | Intercalated cell

K^+
ATP
H^+

$NH_3 + H^+$
+
PO_4

H^+
ATP
H_2CO_3
CA
$CO_2 + H_2O$

Cl^-
HCO_3^-

NH_4^+

Acid excreted

CA, carbonic anhydrase
PO_4, phosphate

The pH of arterial blood is 7.35–7.45 ([H^+]=45–35 nmol L^{-1}). Metabolism produces ~60 mmol H^+ per day, most of which is excreted through the lungs as CO_2, formed by the reaction of H^+ with HCO_3^- (bicarbonate) (Fig. 32a). The kidneys conserve and replace HCO_3^- lost in this way, and fine tune H^+ excretion. Physiological **buffers** maintain a low *free* [H^+] and prevent large swings in pH.

Buffers

Buffers are **weak acids** (HA) or **bases** (A$^-$) that can donate or accept H^+ ions. The ratio between buffer pairs (e.g. carbonic acid, H_2CO_3, and bicarbonate, HCO_3^-) is determined by [H^+] and the **dissociation constant** (K) for that buffer pair: K=([H^+][A$^-$])/[HA], or pH=pK+log([A$^-$]/[HA]) (the **Henderson–Hasselbalch equation**). Thus, an increase in [A$^-$] or a decrease in [HA] will increase pH (more alkaline), and a decrease in pH will decrease the ratio [A$^-$]/[HA]. Buffers work best when the pH is close to their **pK value**, the pH at which the ratio [A$^-$]/[HA] is unity. Bicarbonate and carbonic acid (formed by the combination of CO_2 with water; Fig. 32a) are the most important buffer pair in the body, although haemoglobin provides ~20% of buffering in the blood; phosphate and proteins provide intracellular buffering. Buffers in urine, largely phosphate, allow the excretion of large quantities of H^+.

Although the HCO_3^- system has a pK value of 6.1, and is theoretically a poor buffer at pH 7.4, it is physiologically effective because CO_2 (and therefore H_2CO_3) and HCO_3^- are precisely controlled by the lungs (Chapter 25) and kidney, respectively. These fix the HCO_3^-/H_2CO_3 ratio and therefore the pH, and the latter determines the ratio of all other buffer pairs. The relationship between pH, $P\text{CO}_2$ and [HCO_3^-] is described in Fig. 32a,b. The line 'BAC' is the **buffer line** for whole blood; changes in $P\text{CO}_2$ alter HCO_3^- and pH along this line. Point A denotes normal conditions (pH 7.4, [HCO_3^-]=24 mM, $P\text{CO}_2$=5.3 kPa).

Proximal renal tubule

Bicarbonate is freely filtered, and so filtrate [HCO_3^-] is ~24 mmol L^{-1} (as in plasma). Less than 0.1% of filtered HCO_3^- is normally excreted in the urine, ~80% being reabsorbed in the proximal tubule. HCO_3^- is not transported directly. Filtered HCO_3^- associates with H^+ secreted by epithelial **Na$^+$–H$^+$ antiporters** to form H_2CO_3, which rapidly dissociates to CO_2 and H_2O in the presence of **carbonic anhydrase**. CO_2 and H_2O diffuse into the tubular cells, where they recombine into H_2CO_3, which dissociates to H^+ and HCO_3^-. HCO_3^- is transported into the interstitium largely by **Na$^+$–HCO$_3^-$ symporters** (Fig. 32c). For each H^+ secreted into the lumen, one HCO_3^- and one Na$^+$ enter the plasma. H^+ is recycled, so that there is little net H^+ secretion at this stage. A further 10–15% of HCO_3^- is similarly reabsorbed in the thick ascending loop of Henle. When plasma and thus filtrate [HCO_3^-] becomes greater than ~27 mmol L^{-1}, the reabsorption mechanisms saturate and HCO_3^- is excreted in the urine.

Ammonia is produced in tubular cells by the metabolism of glutamine, which leads to the generation of HCO_3^- and glucose or CO_2. NH_3 diffuses into the tubular fluid, or as NH_4^+ is transported by the Na$^+$–H$^+$ antiporter. In the tubular fluid, NH_3 gains H^+ to form NH_4^+, which cannot diffuse through membranes (Fig. 32d). About 50% of NH_4^+ secreted by the proximal tubule is reabsorbed in the thick ascending loop of Henle, where it substitutes for K$^+$ in the Na$^+$–K$^+$–2Cl$^-$ symporter (Chapter 30), and passes into the medullary interstitium (Fig. 32e). Here, NH_4^+ dissociates into NH_3 and H^+, and NH_3 re-enters the collecting duct by diffusion. The secretion of H^+ in the collecting duct (see below) leads to conversion back to NH_4^+, which is trapped in the lumen and excreted.

Distal renal tubule

The secretion of H^+ in the distal tubule promotes the reabsorption of any remaining HCO_3^-. The combination of H^+ with NH_3 (see above) and phosphate prevents H^+ recycling and allows acid excretion. In the early distal nephron, H^+ secretion is predominantly by Na$^+$–H$^+$ exchange, but more distally secretion is via **H$^+$-ATPase** and **H$^+$-K$^+$-ATPase** in **intercalated cells**, which contain plentiful carbonic anhydrase. As secreted H^+ is derived from CO_2, HCO_3^- is formed and returns to the blood (Fig. 32f).

Summary

In the proximal nephron, H^+ secretion promotes HCO_3^- reabsorption. In the distal nephron, secretion leads to the combination of H^+ with urinary buffers (phosphate, NH_3), and thus the generation of HCO_3^- and acid excretion. As a result of this, tubular fluid becomes more acid as it moves through the nephron. H^+ secretion is proportional to intracellular [H^+], which is itself related to extracellular pH. A fall in blood pH will therefore stimulate renal H^+ secretion.

Acid–base regulation and compensation

Respiratory acidosis and **alkalosis** refer to alterations in pH caused by changes in $P\text{CO}_2$ (i.e. ventilation). **Metabolic acidosis** and **alkalosis** refer to changes not related to $P\text{CO}_2$ (i.e. increased acid production, diet, renal disease). Thus, hypoventilation increases $P\text{CO}_2$ and causes respiratory acidosis, denoted by the move from A to B in Fig. 32b. A *sustained* respiratory acidosis (e.g. *respiratory failure*) can be **compensated** by increased renal excretion of H^+ and reabsorption of HCO_3^-. The [HCO_3^-]/$P\text{CO}_2$ ratio is thus restored, and the pH returns towards normal. This **renal compensation** is denoted by the arrow B to D (Fig. 32b). Similarly, **metabolic acidosis** (G) may be compensated by increased ventilation and reduced $P\text{CO}_2$ (G to E) (**respiratory compensation**), initiated by the detection of acid pH by the chemoreceptors (Chapter 25). Renal mechanisms are slow because their capacity for handling H^+ and HCO_3^- is smaller than that of the lungs for handling CO_2.

K$^+$ homeostasis and acid–base status

Hypokalaemia (low plasma [K$^+$]) is associated with metabolic alkalosis, due to stimulation of ammonia production, Na$^+$–H$^+$ exchange and H$^+$–K$^+$-ATPase, all of which enhance H^+ secretion. This is potentiated by aldosterone (Chapter 31). Hyperkalaemia has the opposite effect, and inhibits NH_4^+ reabsorption by competition at the Na$^+$–K$^+$–2Cl$^-$ symporter. Changes in acid–base status can affect K$^+$ homeostasis for similar reasons.

33 Gastrointestinal tract: overview and the mouth

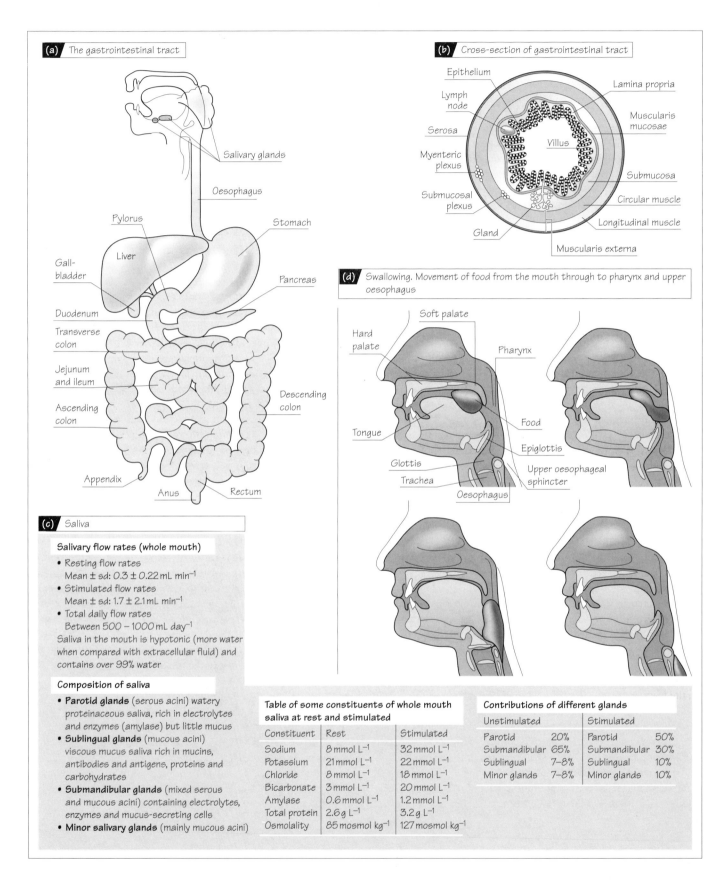

(a) The gastrointestinal tract

Salivary glands
Oesophagus
Pylorus
Stomach
Liver
Gall-bladder
Pancreas
Duodenum
Transverse colon
Jejunum and ileum
Descending colon
Ascending colon
Appendix
Anus
Rectum

(b) Cross-section of gastrointestinal tract

Epithelium
Lymph node
Serosa
Myenteric plexus
Submucosal plexus
Gland
Lamina propria
Muscularis mucosae
Villus
Submucosa
Circular muscle
Longitudinal muscle
Muscularis externa

(d) Swallowing. Movement of food from the mouth through to pharynx and upper oesophagus

Soft palate
Hard palate
Pharynx
Food
Tongue
Epiglottis
Glottis
Upper oesophageal sphincter
Trachea
Oesophagus

(c) Saliva

Salivary flow rates (whole mouth)
- Resting flow rates
  Mean ± sd: 0.3 ± 0.22 mL min⁻¹
- Stimulated flow rates
  Mean ± sd: 1.7 ± 2.1 mL min⁻¹
- Total daily flow rates
  Between 500 – 1000 mL day⁻¹

Saliva in the mouth is hypotonic (more water when compared with extracellular fluid) and contains over 99% water

Composition of saliva
- **Parotid glands** (serous acini) watery proteinaceous saliva, rich in electrolytes and enzymes (amylase) but little mucus
- **Sublingual glands** (mucous acini) viscous mucus saliva rich in mucins, antibodies and antigens, proteins and carbohydrates
- **Submandibular glands** (mixed serous and mucous acini) containing electrolytes, enzymes and mucus-secreting cells
- **Minor salivary glands** (mainly mucous acini)

Table of some constituents of whole mouth saliva at rest and stimulated

Constituent	Rest	Stimulated
Sodium	8 mmol L⁻¹	32 mmol L⁻¹
Potassium	21 mmol L⁻¹	22 mmol L⁻¹
Chloride	8 mmol L⁻¹	18 mmol L⁻¹
Bicarbonate	3 mmol L⁻¹	20 mmol L⁻¹
Amylase	0.6 mmol L⁻¹	1.2 mmol L⁻¹
Total protein	2.6 g L⁻¹	3.2 g L⁻¹
Osmolality	85 mosmol kg⁻¹	127 mosmol kg⁻¹

Contributions of different glands

Unstimulated		Stimulated	
Parotid	20%	Parotid	50%
Submandibular	65%	Submandibular	30%
Sublingual	7–8%	Sublingual	10%
Minor glands	7–8%	Minor glands	10%

The **gastrointestinal (GI) tract** is responsible for the breakdown of food into its component parts so that they can be absorbed into the body. It is made up of the **mouth, oesophagus, stomach** and the **small and large intestines**. The **salivary glands, liver, gallbladder** and **pancreas** are organs distinct from the GI tract, but all secrete juices into the tract and aid the digestion and absorption of the food (Fig. 33a).

Different regions of the tract are concerned with **motility (transport), storage, digestion, absorption** and **elimination of waste**, and these functions of the GI tract are controlled by **neuronal, hormonal** and **local regulatory mechanisms**.

The walls of the GI tract have a general structure that is similar along most of its length, although this is modified as function varies. This basic structure is shown in Fig. 33b. It comprises the **mucosal layer**, made up of epithelial cells (which can be involved in either the process of secretion or absorption depending on their location in the GI tract), and the **lamina propria**, consisting of loose connective tissue, collagen and elastin, blood vessels and lymph tissue, and a thin layer of smooth muscle called the **muscularis mucosa** which, when contracting, produces folds and ridges in the mucosa. The **submucosal layer** comprises a second layer of connective tissue, but also contains larger blood and lymphatic vessels and a network of nerve cells called the **submucosal plexus (Meissner's plexus)**. This is a dense plexus of nerves innervated by the autonomic part of the nervous system which can function as an independent nervous system—the **enteric nervous system**. Below the submucosa is the **muscularis externa**. This comprises a thick **circular layer** of smooth muscle around the GI tract which, when it contracts, produces a constriction of the lumen. Below this layer of muscle is another thinner layer of muscle arranged in a **longitudinal** manner which, when it contracts, results in shortening of the tract. Between these two layers of muscle is a second nerve plexus, called the **myenteric plexus (Auerbach's plexus)**, which is also part of the enteric nervous system. The outermost layer of the GI tract is the **serosa**, another connective tissue layer covered with squamous mesothelial cells.

The GI tract starts in the mouth, where food is initially **chewed (masticated)** and mixed with salivary secretions. **Mastication** is the process of systematic mechanical breakdown of food in the mouth. The amount of mastication necessary in order to swallow the food depends on the nature of the ingested food: solid foods are subjected to vigorous chewing, whereas softer foods and liquids require little or no chewing and are transported almost directly into the oesophagus by swallowing. Mastication is necessary for some foods, such as red meats, chicken and vegetables, to be fully absorbed by the rest of the GI tract. However, fish, eggs, rice, bread and cheese do not require chewing for complete absorption in the tract.

Mastication involves the coordinated activity of the **teeth, jaw muscles, temporomandibular joint, tongue** and other structures, such as the **lips, palate** and **salivary glands**. The forces developed between the teeth during mastication have been measured to be about 150–200 N; however, the maximum biting force developed between the molar teeth is almost 10 times this value.

During mastication three pairs of glands, the **parotid, submandibular** and **sublingual**, secrete saliva. The major functions of saliva are to **moisten** and **lubricate** the mouth at rest, but particularly during eating and speech, to **dissolve** food molecules so that they can react with gustatory receptors giving rise to the sensation of taste, to **ease swallowing**, to begin the early part of **digestion** of polysaccharides (complex sugars) and to **protect** the oral cavity by coating the teeth with a proline-rich protein or pellicle that can serve as a protective barrier on the tooth surface. Saliva also contains immunoglobulins that have a protective role in avoiding bacterial infections.

Saliva is **hypotonic** and contains a mixture of both inorganic and organic constituents. The composition varies according to which gland is secreting and also whether it is resting or being stimulated (Fig. 33c).

The **control of salivary secretion** depends on reflex responses which, in humans, have been shown to be elicited by the stimulation of gustatory (taste) receptors and periodontal and mucosal mechanoreceptors during mastication. Although it was thought that olfactory afferent stimulation (smell) also had a general reflex effect on salivary secretion, it has now been shown that this reflex operates via the submandibular/sublingual glands and not the parotid in humans. The sight and thought of food in humans have very little effect on salivary production. The perception of an increased salivary production is thought to be related to the sudden awareness of saliva already present in the mouth.

Swallowing occurs in a number of phases. The first phase is **voluntary** and involves the formation of a bolus of food by chewing and tongue movements (backwards and upwards) which push the food into the pharynx. The remaining phases are not voluntary, but **reflex responses** initiated by the stimulation of mechanoreceptors with afferents in the **glossopharyngeal (IX) and vagus (X) nerves** to the medulla and pons (brain stem); here, there is a group of neurones ('**the swallowing centre**') which coordinates the complex sequence of events that eventually deliver the bolus into the oesophagus. The **soft palate** elevates to prevent food from entering the **nasal cavity**, respiration is inhibited, the **larynx** is raised, the **glottis** is closed and the food pushes the tip of the **epiglottis** over the tracheal opening, preventing food from entering the trachea. As the bolus enters the **oesophagus**, these changes reverse, the larynx opens and breathing continues (Fig. 33d).

34 Oesophagus and stomach

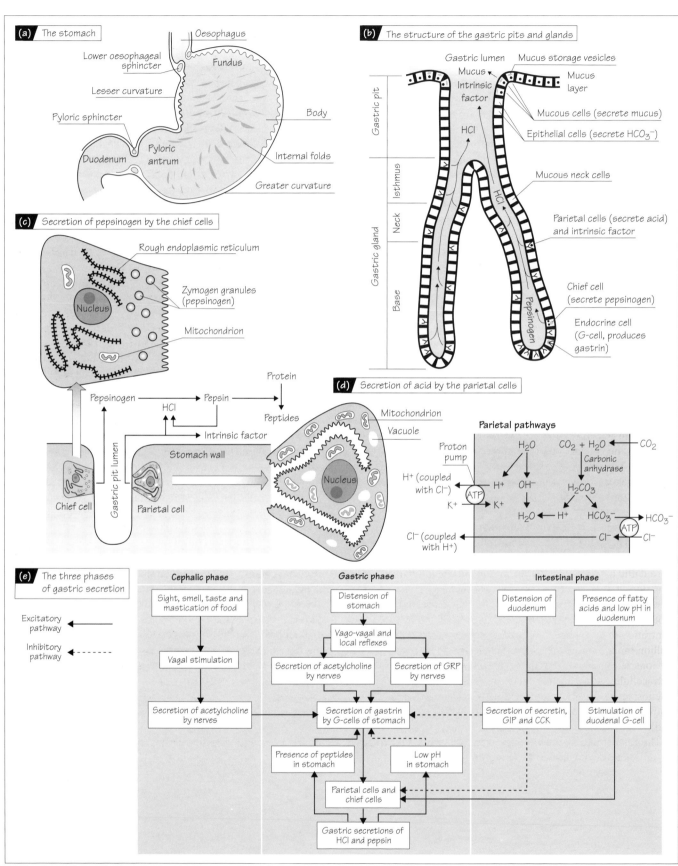

(a) The stomach

Oesophagus
Lower oesophageal sphincter
Lesser curvature
Pyloric sphincter
Duodenum
Pyloric antrum
Fundus
Body
Internal folds
Greater curvature

(b) The structure of the gastric pits and glands

Gastric lumen
Mucus
Intrinsic factor
HCl
Mucus storage vesicles
Mucus layer
Mucous cells (secrete mucus)
Epithelial cells (secrete HCO_3^-)
Mucous neck cells
Parietal cells (secrete acid) and intrinsic factor
Chief cell (secrete pepsinogen)
Endocrine cell (G-cell, produces gastrin)
Gastric pit
Isthmus
Neck
Base
Gastric gland
Pepsinogen

(c) Secretion of pepsinogen by the chief cells

Rough endoplasmic reticulum
Zymogen granules (pepsinogen)
Nucleus
Mitochondrion
Pepsinogen → Pepsin → Protein → Peptides
HCl
Intrinsic factor
Chief cell
Gastric pit lumen
Stomach wall
Parietal cell

(d) Secretion of acid by the parietal cells

Mitochondrion
Vacuole
Nucleus
Parietal pathways
Proton pump
H_2O
$CO_2 + H_2O$ ← CO_2
Carbonic anhydrase
H^+ (coupled with Cl^-)
H^+
OH^-
H_2CO_3
K^+
K^+
H_2O ← H^+
HCO_3^- → HCO_3^-
Cl^- (coupled with H^+)
Cl^- ← Cl^-
ATP

(e) The three phases of gastric secretion

Excitatory pathway →
Inhibitory pathway ⇠

Cephalic phase

Sight, smell, taste and mastication of food
↓
Vagal stimulation
↓
Secretion of acetylcholine by nerves

Gastric phase

Distension of stomach
↓
Vago-vagal and local reflexes
↓
Secretion of acetylcholine by nerves | Secretion of GRP by nerves
↓
Secretion of gastrin by G-cells of stomach
Presence of peptides in stomach | Low pH in stomach
↓
Parietal cells and chief cells
↓
Gastric secretions of HCl and pepsin

Intestinal phase

Distension of duodenum | Presence of fatty acids and low pH in duodenum
↓
Secretion of secretin, GIP and CCK | Stimulation of duodenal G-cell

It is possible to swallow food and drink and for it to enter the stomach whilst standing on one's head or experiencing zero gravity. A ring of skeletal muscle called the **upper oesophageal sphincter** usually closes the **pharyngeal end** of the **oesophagus**. During the **oesophageal phase of swallowing**, this sphincter is relaxed, allowing the bolus of food to pass through it. Immediately afterwards, the sphincter closes. Once in the oesophagus, the bolus is propelled the 25 cm (approximately) to the **stomach** by a process called **peristalsis**, a coordinated wave of relaxation in front of the bolus and contraction behind the bolus of the circular and longitudinal muscle layers of the oesophagus, forcing the food into the stomach in about 5 s. Before the bolus enters the stomach, it passes through another sphincter, the **lower oesophageal sphincter**, formed from a ring of smooth muscle which relaxes as the peristaltic wave reaches it. The **swallowing centres in the medulla** produce a sequence of events that lead to both efferent activity to **somatic nerves** (innervating skeletal muscle) and **autonomic nerves** (innervating smooth muscle). This sequence of events is influenced by afferent receptors in the oesophagus wall sending impulses back to the medulla. The sphincters and the peristaltic waves are principally controlled by activity in the **vagus nerve** and aided by a high degree of coordination of the activity within the **enteric nerve plexuses** within the tract itself.

Once the bolus of food passes through the lower oesophageal sphincter, it enters the **stomach** (Fig. 34a). The **main functions of the stomach** are to **store** food temporarily (as it can be ingested more rapidly than it can be digested) to chemically and mechanically **digest** food using acids, enzymes and movements, to **regulate the release** of the resulting **chyme** into the small intestine, and to secrete a substance called **intrinsic factor** which is essential for the absorption of vitamin B_{12}. The stomach lies immediately below the diaphragm and, like the rest of the gastrointestinal tract, it has longitudinal and circular muscle layers and nerve plexuses in its walls; however, within the mucosa are specialized secretory cells that line the gastric glands or pits (Fig. 34b). When empty, the stomach has a volume of approximately 50 mL; however, when fully distended, its volume can be as much as 4 L. **Proteins** in the food are broken down into **polypeptides** in the stomach by enzymes called **pepsins**. These enzymes are produced in an inactive form called **pepsinogens** by the **chief cells** in the gastric mucosa, and are converted into active pepsins by the acid environment in the stomach (Fig. 34c). The acid in the stomach is **hydrochloric acid** and is produced by a specialized group of cells called **parietal cells**. The stomach can secrete as much as 2 L of acid per day, and the concentration of H^+ ions in the stomach is estimated to be about one million times higher than that in the blood. This concentration of H^+ ions requires a very efficient exchange of intracellular H^+ for extracellular K^+ using energy provided by the breakdown of adenosine triphosphate (ATP). This is achieved using a protein known as the proton pump or the H^+–K^+-ATPase protein (Fig. 34d).

The gastric mucosa does not digest itself because it is protected by an **alkaline, mucin-rich fluid** secreted by the gastric glands, which acts as a mucosal barrier by bathing the gastric epithelial cells. In addition, local mediators, such as **prostaglandins**, are released when the mucosa is irritated, and these increase the thickness of the **mucus layer** and stimulate the production of **bicarbonate** which neutralizes the acid.

Gastric secretions occur in basically three phases: **cephalic, gastric** and **intestinal** (Fig. 34e). The **cephalic phase** is brought about by the **sight, smell, taste** and **mastication** of food. At this stage, there is no food in the stomach and acid secretion is stimulated by the activation of the vagus and its actions on the **enteric plexus**. **Postganglionic parasympathetic fibres** in the **myenteric plexus** cause the release of **acetylcholine (ACh)** and stimulate the release of gastric juices from the gastric glands. Vagal stimulation also causes the release of a hormone called gastrin from cells in the antrum of the stomach called **G-cells**. **Gastrin** is secreted into the bloodstream and, when it reaches the gastric glands, it stimulates the release of **acid** and **pepsinogens**. Both vagal activity and gastrin also stimulate the release of **histamine** from mast cells, which, in turn, acts on **parietal cells** to produce more acid.

When food arrives in the stomach, it stimulates the **gastric phase** of secretion of **acid, pepsinogen** and **mucus**. The main stimuli for this phase are the **distension** of the stomach and the **chemical** composition of the food. Mechanoreceptors in the stomach wall are stretched and set up local myenteric reflexes and also longer vago-vagal reflexes. Both cause the release of **ACh** which stimulates the release of **gastrin, histamine** and, in turn, **acid, enzymes** and **mucus**. Stimulation of the **vagus** also releases a specific peptide, **gastrin-releasing peptide (GRP)**, which mainly acts directly on the **G-cells** to release **gastrin**. Whole proteins do not affect gastric secretions directly, but their breakdown products, such as **peptides** and **free amino acids**, do so by **directly stimulating gastrin secretion**. A **low pH** (more acid) in the stomach **inhibits** gastrin secretion; therefore, when the stomach is empty or after food has entered it and acid has been secreted for some time, there is an inhibition of acid production. However, when food first enters the stomach, the **pH rises** (less acid) and this leads to a **release of the inhibition** and causes a **maximum secretion of gastrin**. Thus, gastric acid secretion is **self-regulating**.

The **gastric phase** normally lasts for about 3 h and the food in the stomach is converted into a sludge-like material called **chyme**. The **chyme** enters the first part of the small intestine, the **duodenum**, through the **pyloric sphincter**. The presence of chyme in the **pyloric antrum** distends it and causes antral contractions and opening of the sphincter. The rate at which the stomach empties depends on the volume in the antrum and the fall in the pH of the chyme, both leading to an increase in emptying. However, **distension** of the duodenum, the **presence of fats** and a **decrease in pH** in the duodenal lumen all cause **an inhibition of gastric emptying**. This mechanism leads to a precise supply of chyme to the intestines at a rate appropriate for it to be digested properly.

(For a description of the intestinal phase of gastric secretion, see Chapter 35).

35 Small intestine

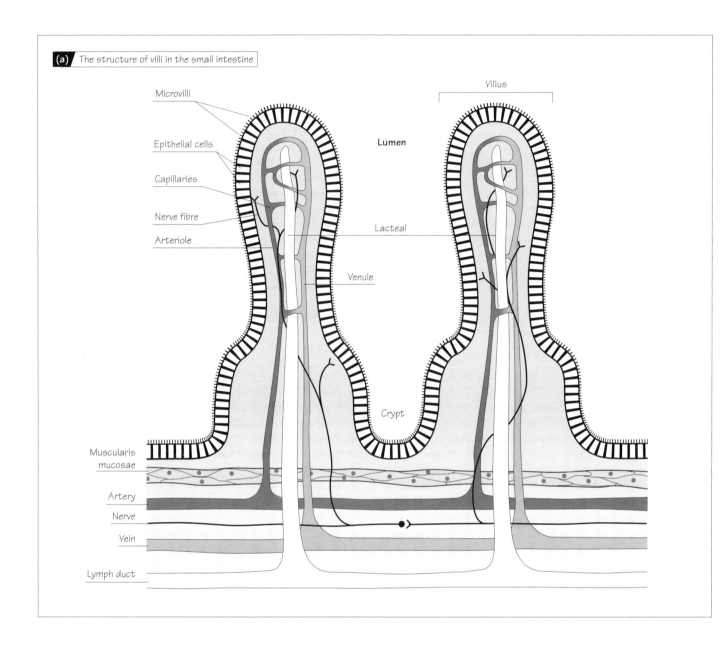

(a) The structure of villi in the small intestine

Microvilli

Epithelial cells

Capillaries

Nerve fibre

Arteriole

Villus

Lumen

Lacteal

Venule

Crypt

Muscularis mucosae

Artery

Nerve

Vein

Lymph duct

The small intestine is the main site for the digestion of food and the absorption of the products of this digestion. It is a tube, 2.5 cm in diameter and approximately 4 m in length, and comprises the **duodenum**, **jejunum** and **ileum**.

When **chyme** first enters the duodenum, there is a continuation of gastric secretion thought to be due to the activation of **G cells** in the **intestinal mucosa** (see **intestinal phase**; Fig. 34e). This is short lived as the duodenum becomes more distended with further gastric emptying. A series of reflexes is initiated which inhibits the further release of **gastric juices**. A number of **hormones** are involved in these reflex responses. **Secretin** is released in response to acid stimulation; it reaches the stomach via the bloodstream and inhibits the release of **gastrin**. The presence of **fatty acids**, due to the breakdown of fats in the duodenum

itself, releases two polypeptide hormones, called **gastric inhibitory peptide (GIP)** and **cholecystokinin (CCK)**, which inhibit the release of both **gastrin** and **acid**. Both secretin and CCK, however, stimulate the release of **pepsinogen** from the **chief cells**, thereby aiding protein digestion. Together with **mechanoreceptors** in the duodenum via vagal and local reflex pathways, the release of **secretin** and **CCK** has also been implicated in the control of **gastric emptying**. The chyme that first enters the duodenum is **acidic**, **hypertonic** and **only partly digested**; at this early stage, the nutrients formed cannot be absorbed. There is an **osmotic** movement of water across the freely permeable wall which leads to the contents becoming **isotonic**. The acidity is neutralized by the addition of both **bicarbonate** secreted by the **pancreas** and **bile** from the **liver**, and further digestion of the chyme is performed by the addition of enzymes from the pancreas, liver and intestine itself.

The lining of the small intestine is folded into many small, finger-like projections called **villi** (Fig. 35). Between the villi lie some small glands, called **crypts**, which can secrete up to 3 L of **hypotonic** fluid per day. The surface of the villi is covered with a layer of **epithelial cells** which, in turn, have many small projections called **microvilli** (collectively called the **brush-border**) that project towards the lumen of the intestine. The small intestine is particularly adapted for the absorption of nutrients. It has a huge surface area (about the size of a tennis court), and the chyme is forced into a circular motion as it passes through the tract, facilitating mixing and therefore digestion and absorption. There is a constant turnover of epithelial cells within the gastrointestinal (GI) tract, with the small intestine epithelium totally replacing itself approximately every 6 days.

Each **villus** contains a single, blind-ended lymphatic vessel, called a **lacteal**, and also a **capillary network**. Most nutrients are absorbed into the bloodstream via these vessels. The venous drainage from the small intestine, large intestine, pancreas and also from some parts of the stomach passes via the **hepatic portal vein** into the liver; here, it passes through a second capillary bed to be further processed before returning to the circulation.

The small intestine absorbs **water, electrolytes, carbohydrates, amino acids, minerals, fats** and **vitamins**. The mechanisms by which movement from the lumen to the circulation occurs are variable. Nutrients move between the GI tract and the blood by passing through and around the epithelial cells. As the contents of the intestine are isotonic with body fluids and mostly have the same concentration of the major electrolytes, their absorption is active. **Water** cannot be moved directly, but follows osmotic gradients set up by the transport of ions. The major contributor to this osmotic gradient is the **sodium pump**. Na^+–K^+-ATPase is located on the blood side of the epithelial cell (**basolateral membrane**), and hydrolysis of adenosine triphosphate (ATP) to adenosine diphosphate (ADP) leads to the expulsion of three Na^+ ions from the cell in exchange for two K^+ ions. Both of these are against the concentration gradients, leading to a **low concentration of Na^+** and a **high concentration of K^+** within the cells. The low intracellular concentration of Na^+ ensures a movement of Na^+ from the intestinal contents into the cell by both **membrane channels** and **transported protein mechanisms**. Na^+ is then rapidly transported out of the cell again by the **basolateral Na^+–K^+ pump**. K^+ leaves the cell, again via the basolateral membrane, down its concentration gradient. This outward movement of K^+ is linked to an outward movement of Cl^-, against its concentration gradient, Cl^- having entered down its concentration gradient like Na^+ via the lumenal membrane. These movements set up an osmotic gradient between the lumen and the blood, leading to water absorption following the movement of Na^+ and Cl^- from the lumen into the cell across the lumenal membrane.

Carbohydrates are absorbed mostly in the form of **monosaccharides (glucose, fructose** and **galactose**). They are broken down into monosaccharides by enzymes released from the brush-border (**maltases, isomaltases, sucrase** and **lactase**). The monosaccharides are transported across the epithelium into the bloodstream by means of cotransporter molecules that link their inward movement with that of Na^+ down its concentration gradient. At the basolateral membrane, monosaccharides leave the cell either by **simple diffusion** or by **facilitated diffusion** down the concentration gradient.

The **polypeptides** produced in the stomach are broken down into **oligopeptides** in the small intestine by enzymes (**proteases**) secreted by the pancreas: **trypsin** and **chymotrypsin**. These are further broken down into **amino acids** by another pancreatic enzyme, called **carboxypeptidase**, and an enzyme located on the lumenal membrane epithelial cells, called **aminopeptidase**. The **free amino acids** enter the epithelial cells by secondary active transport coupled to the movement of Na^+ and a number of different cotransporter mechanisms.

Two very important minerals that are absorbed from the diet are **calcium** and **iron**. **Intracellular calcium** concentrations are low and any **free calcium** in the diet can cross the lumenal membrane down a steep concentration gradient through channels or by a carrier mechanism. In the cell, it binds to a protein which carries it to the basolateral membrane, where it is actively transported against the concentration gradient by a **Ca^{2+}-ATPase** with the hydrolysis of ATP, or by an **Na^+–Ca^{2+} antiporter** linked with the movement of Na^+ down its concentration gradient into the cell and the removal of Ca^{2+} from it.

Most **dietary iron** is in the **ferric** form which **cannot** be absorbed; however, in the **ferrous** form, it forms soluble complexes with **ascorbate** and other substances and **can be readily absorbed**. These complexes are transported across the membrane by a carrier protein and, once in the cell, bind with a variety of substances including **ferritin**. A second carrier protein transports the iron across the basolateral membrane into the bloodstream.

Fat digestion occurs almost entirely in the small intestine. The major enzyme is a **pancreatic enzyme** called **lipase** which breaks fat down into **monoglycerides** and **free fatty acids**. However, before the fat can be broken down, it has to be **emulsified**, which is a process by which the larger lipid droplets are broken down into much smaller droplets (about 1 μm in diameter). The main emulsifying agents are the **bile acids, cholic acid** and **chenodeoxycholic acid**. The free fatty acids and monoglycerides form tiny particles (4–5 nm in diameter) with the bile acids, called **micelles**. The outer region of the micelle is **hydrophilic** (water-attracting), whereas the inner core contains the **hydrophobic** (water-repelling) part of the molecule. This arrangement allows the micelles to enter the aqueous layers surrounding the **microvilli**, and the **monoglycerides, free fatty acids, cholesterol** and **fat-soluble vitamins** can then diffuse passively into the duodenal cells, leaving the bile salts within the lumen of the gut until they reach the ileum, where they are reabsorbed. Once within the epithelial cells, the fatty acids and monoglycerides are reassembled into fats by a number of different metabolic pathways. They then enter the lymphatic system via the **lacteals** and eventually reach the bloodstream through the **thoracic duct**.

The **fat-soluble vitamins**, A, D, E and K, essentially follow the pathways for fat absorption. The remaining **water-soluble vitamins** are mainly absorbed by diffusion or mediated transport. The exception is **vitamin B_{12}**, which must first bind with **intrinsic factor** (secreted from the parietal cells in the stomach wall). When bound, vitamin B_{12} attaches to specific sites on the epithelial cells in the **ileum** where a process of endocytosis leads to absorption.

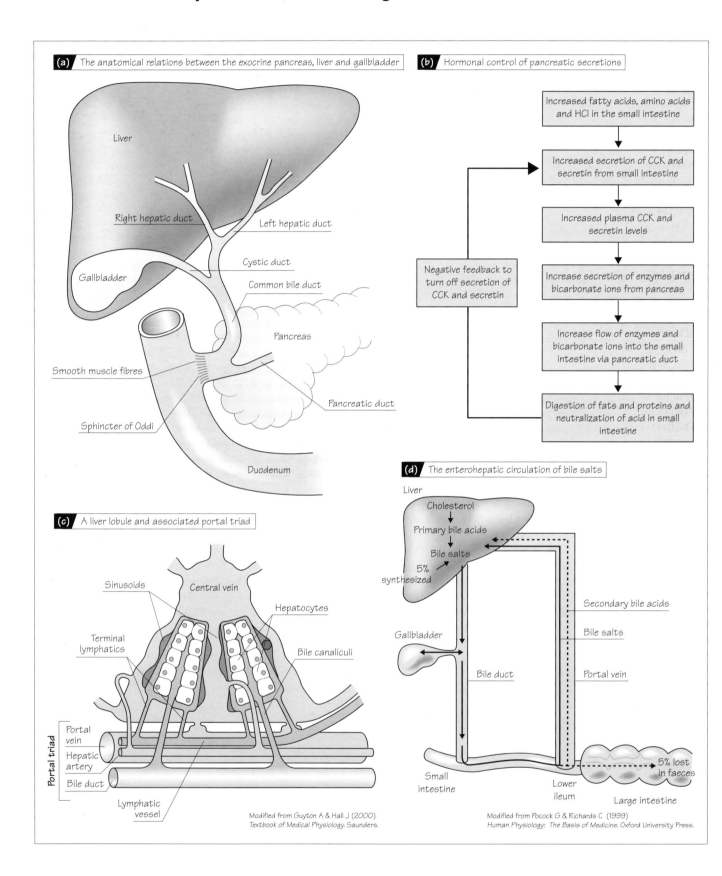

(a) The anatomical relations between the exocrine pancreas, liver and gallbladder

Liver

Right hepatic duct

Left hepatic duct

Cystic duct

Gallbladder

Common bile duct

Pancreas

Smooth muscle fibres

Pancreatic duct

Sphincter of Oddi

Duodenum

(b) Hormonal control of pancreatic secretions

Increased fatty acids, amino acids and HCl in the small intestine

Increased secretion of CCK and secretin from small intestine

Increased plasma CCK and secretin levels

Increase secretion of enzymes and bicarbonate ions from pancreas

Increase flow of enzymes and bicarbonate ions into the small intestine via pancreatic duct

Digestion of fats and proteins and neutralization of acid in small intestine

Negative feedback to turn off secretion of CCK and secretin

(c) A liver lobule and associated portal triad

Sinusoids

Central vein

Hepatocytes

Terminal lymphatics

Bile canaliculi

Portal vein

Hepatic artery

Bile duct

Portal triad

Lymphatic vessel

Modified from Guyton A & Hall J (2000)
Textbook of Medical Physiology. Saunders.

(d) The enterohepatic circulation of bile salts

Liver

Cholesterol

Primary bile acids

Bile salts

5% synthesized

Gallbladder

Bile duct

Small intestine

Lower ileum

Large intestine

Secondary bile acids

Bile salts

Portal vein

5% lost in faeces

Modified from Pocock G & Richards C (1999)
Human Physiology: The Basis of Medicine. Oxford University Press.

The **exocrine pancreas** secretes a major digestive fluid called **pancreatic juice**. This juice is secreted into the duodenum via the pancreatic duct that opens into the gastrointestinal (GI) tract at the same site as the **common bile duct** (see later). When food is present in the duodenum, a small sphincter (**sphincter of Oddi**) relaxes, allowing both bile and pancreatic secretions to enter the tract (Fig. 36a).

Pancreatic juice is made up of a number of **enzymes**, secreted by the **acinar** cells of the pancreas, which break down the major constituents in the diet. The enzymes include **pancreatic amylase**, which breaks down carbohydrates to monosaccharides, **pancreatic lipase**, which breaks down fats to glycerol and fatty acids, **ribonuclease** and **deoxyribonuclease**, which are involved in the breakdown of nucleic acids and free mononucleotides, and a variety of **proteolytic enzymes (trypsin, chymotrypsin, elastase** and **carboxypeptidase)**, which break down proteins into small peptides and amino acids. The hormone **cholecystokinin (CCK)**, released into the bloodstream by the duodenal cells in response to the presence of amino acids and fatty acids in the chyme, is responsible for the secretion of the pancreatic enzymes from the acinar cells of the pancreas. The other major secretions, besides the enzymes, are **water** and **bicarbonate ions**. The volume of pancreatic juice secreted precisely neutralizes the acid content of the chyme delivered by the stomach to the intestines. This is caused by the acid in the duodenum releasing **secretin** from its walls into the bloodstream. **Secretin** stimulates the production of water and bicarbonate ions from the duct system and, in particular, from the **epithelial cells** lining the duct. Approximately 1 L of pancreatic juice is secreted per day from a normal individual (Fig. 36b).

The **liver** is the largest organ of the body, weighing over 1 kg in the normal adult. The **functions of the liver** can be divided into two broad categories. Firstly, it is involved with the processing of absorbed substances, both nutrient and toxic. In other words, it is responsible for the **metabolism** of a vast range of substances produced by the digestion and absorption of food from the intestine. Secondly, it has an important **exocrine function** in that it is involved in: (i) the production of bile acids and alkaline fluids used in the digestion and absorption of fats and for the neutralization of gastric acid in the intestines; (ii) the breakdown and production of waste products following digestion; (iii) the detoxification of noxious substances; and (iv) the excretion of waste products and the detoxification of substances in bile.

The majority of waste metabolites and detoxified substances are excreted from the body, in the bile, from the GI tract, or via secretions from the liver into the bloodstream for subsequent excretion by the kidney. The relationship between the liver, gallbladder and duodenum is shown in Fig. 36a. The **liver** consists of four lobes, with each lobe made up of tens of thousands of hexagonal **lobules**, 1–2 mm in diameter, which are the functional unit of the liver.

Each lobule (Fig. 36c) consists of a **central vein** that eventually becomes part of the **hepatic vein**. Surrounding the central vein are single columns of liver cells (**hepatocytes**) radiating outwards; between the hepatocytes are small **canaliculi** which begin as blind-ended structures at the end nearer the central vein, but drain into the **bile duct** on the periphery of the lobule. At each of the six corners of the lobules lies a 'portal triad' comprising branches of the **hepatic artery**, the **portal vein** and the **bile duct**. The bile ducts eventually drain into the **terminal bile duct**.

The **hepatocytes** secrete a fluid called **hepatic bile**. It is isotonic and resembles plasma ionically. It also contains **bile salts**, **bile pigments**, **cholesterol**, **lecithin** and **mucus**. This fraction of bile is called the **bile acid-dependent fraction**. As it passes along the bile duct, the bile is modified by the epithelial cells lining the duct by the addition of **water** and **bicarbonate ions**; this fraction is called the **bile acid-independent fraction**. Overall, the liver can produce 500–1000 mL of bile per day. The bile is either discharged directly into the duodenum or stored in the **gallbladder**. The bile acid-independent fraction is made at the time it is required, i.e. during digestion of the chyme. The bile acid-dependent fraction is made when the bile salts are returned from the GI tract to the liver, and is then stored in the gallbladder when the sphincter of Oddi is closed. About 95% of the bile salts that enter the small intestine in bile are recycled and reabsorbed into the portal circulation by active transport mechanisms in the distal ileum (the so-called **enterohepatic circulation**) (Fig. 36d). Many of the bile salts are returned unaltered, some are broken down by intestinal bacteria into **secondary bile acids** and then reabsorbed, and a small proportion escapes reabsorption and is excreted in the faeces.

The **gallbladder** not only stores the bile, but also concentrates it by removing non-essential solutes and water, leaving the bile acids and pigments. The process of concentration is mainly by active transport of sodium ions into the intercellular spaces of the lining cells and this, in turn, draws water, bicarbonate and chloride ions from the bile and into the extracellular fluid, thereby concentrating the bile held in the gallbladder.

The formation of bile is stimulated by **bile salts**, **secretin**, **glucagons** and **gastrin**. The release of bile stored in the gallbladder, however, is stimulated by the secretion of **CCK** into the bloodstream when chyme enters the duodenum and, to a lesser extent, by the actions of the **vagus nerve**. Within a few minutes of a meal, particularly when fats are consumed, the muscles of the gallbladder contract; this forces the contents into the duodenum through the now relaxed sphincter of Oddi. CCK relaxes the sphincter and stimulates the pancreatic secretions at the same time. The gallbladder empties completely 1 h after a fat-rich meal and maintains the level of bile acids in the duodenum above that necessary for the function of the micelles.

37 Large intestine

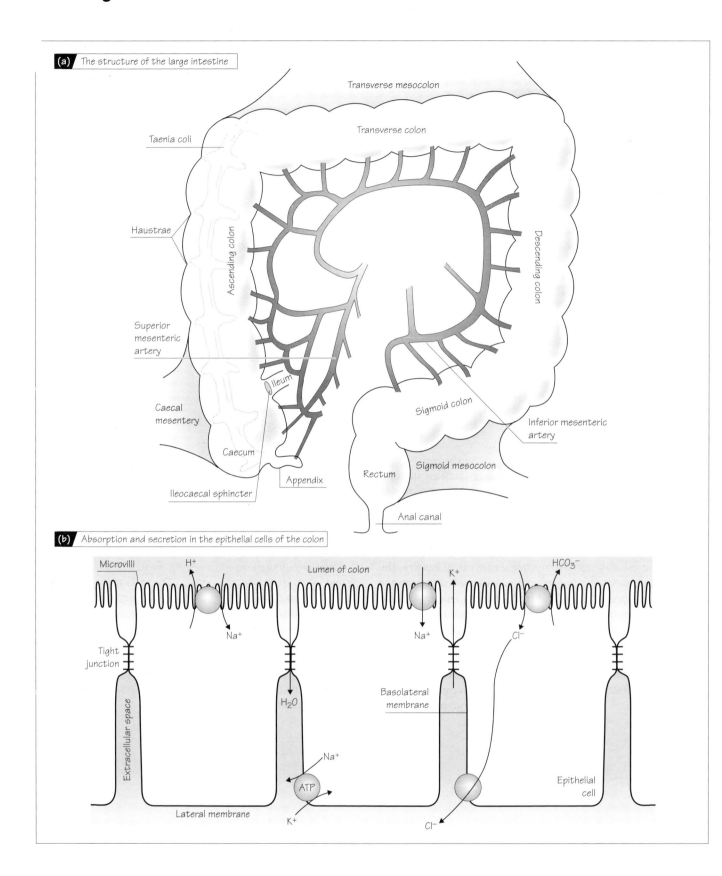

(a) The structure of the large intestine

Transverse mesocolon

Transverse colon

Taenia coli

Haustrae

Ascending colon

Descending colon

Superior mesenteric artery

Ileum

Caecal mesentery

Sigmoid colon

Inferior mesenteric artery

Caecum

Sigmoid mesocolon

Rectum

Ileocaecal sphincter

Appendix

Anal canal

(b) Absorption and secretion in the epithelial cells of the colon

Microvilli

H^+

Lumen of colon

K^+

HCO_3^-

Na^+

Na^+

Cl^-

Tight junction

H_2O

Basolateral membrane

Extracellular space

Na^+

ATP

Epithelial cell

Lateral membrane

K^+

Cl^-

The **large intestine** comprises the **caecum, ascending, transverse, descending** and **sigmoid colon, rectum** and **anal canal** (Fig. 37a). It is approximately 1.2 m in length and between 6 and 9 cm in diameter. Approximately 1.5 L of chyme enters the large intestine per day through a sphincter called the **ileocaecal sphincter**. Distension of the terminal ileum results in the opening of the sphincter and distension of the caecum causes it to close, thereby maintaining the optimum rate of entry to maximize the main function of the large intestine, which is to absorb most of the water and electrolytes. The initial 1.5 L is reduced to about 150 g of **faeces** consisting of 100 mL of water and 50 g of solids.

The muscle layers of the large intestine are slightly different from those found in the rest of the gastrointestinal (GI) tract. It still has a powerful circular muscle layer, but its **longitudinal muscle layer** is concentrated into **three bands** called the **taeniae coli**. The caecum and the ascending and transverse colon are innervated by **parasympathetic** branches of the **vagus**; the descending and sigmoid colon, rectum and anal canal are innervated by **parasympathetic** branches of the **pelvic nerves** from the sacral spinal cord. These parasympathetic fibres innervate the intramural plexuses. The **sympathetic** nerves via the superior mesenteric plexus, and via the inferior mesenteric plexus and the superior hypogastric plexuses, innervate the proximal and distal parts of the large intestine, respectively. The rectum and anal canal are innervated via the inferior hypogastric plexus. Stimulation of the parasympathetic fibres causes segmental contraction, whereas stimulation of the sympathetic fibres stops colonic activity. The **internal** and **external anal sphincters** usually keep the anal canal closed and are controlled both **reflexly** and **voluntarily**. The **internal sphincter** is made up of **circular smooth muscle**, and the more distal **external sphincter** is composed of **striated muscle** which is innervated by motor fibres from the **pudendal nerve**.

Movement of the chyme through the large intestine involves both **mixing** and **propulsion**. However, as the main function is to store the residues of food and to absorb water and electrolytes from it, the movements are slow and sluggish (approximately 5–10 cm h^{-1}). Chyme usually remains in the colon for up to 20 h. The mixing movement is called **haustration** and the sac-like compartments in the colon caused by this process are called **haustra**. The contents of the haustra are often shunted back and forth from one to another in a process called **haustral shuttling**. This aids the exposure of chyme to the mucosal surface and helps the reabsorption of water and electrolytes. In the distal parts of the colon, the contractions are slower and less propulsive, and eventually the faeces collect in the descending colon.

Several times a day there is an increase in activity within the colon, in which there is a vigorous propulsive movement, the **mass movement**. This results in the emptying of a large proportion of the content of the proximal colon into the more distal parts. This **mass movement** is initiated by a complex series of intrinsic reflex pathways started by the distension of the stomach and duodenum soon after the consumption of a meal.

When a critical mass of faeces is forced into the rectum, the desire for **defecation** is experienced. This sudden distension of the rectum walls produced by the final mass movement leads to a **defecation reflex**. This reflex comprises a contraction of the rectum, relaxation of the internal anal sphincter and, initially, contraction of the external anal sphincter. This initial contraction is soon followed by a reflex relaxation of the sphincter initiated by an increase in the peristaltic activity in the sigmoid colon and pressure in the rectum. The faeces are then expelled. This reflex relaxation can be overridden by higher centre activity, leading to a voluntary control over the sphincter which can delay the expelling of faeces. The prolonged distension of the rectum then leads to a **reverse peristalsis**, which empties the rectum into the colon and removes the urge to defecate until the next mass movement and/or a more convenient time.

The chyme that enters the large intestine is **isotonic**; however, in the colon more water than electrolytes is absorbed, leading to water being absorbed against a concentration gradient. The process is controlled by **Na$^+$–K$^+$-ATPases** located in the basolateral and lateral membranes of the epithelial cells that line the walls (Fig. 37b). The mucosal surface of the large intestine is relatively smooth with no villi (only microvilli); however, crypts are present and the majority of cells are **columnar** absorptive cells with a large number of mucus-secreting goblet cells. Na$^+$ is extruded by the membrane pumps into the extracellular spaces. **Tight junctions** at the lumenal side of the cells prevent the diffusion of Na$^+$ and Cl$^-$ from the extracellular spaces into the lumen; this leaves a hypertonic solution close to the lumen, causing water to diffuse from the contents of the lumen. The electrolytes are absorbed by a variety of mechanisms similar to those described for the small intestine. Essentially, there is a net movement of K$^+$ and bicarbonate ions from the blood into the large intestine because of the potential difference set up by the asymmetrical absorption of Na$^+$ and Cl$^-$ across the cell wall.

The majority of the **bacteria** that are present in the GI tract are found in the large intestine, because the acid environment in the rest of the tract destroys most of the so-called microflora. Ninety-nine per cent of the bacteria are anaerobic and most are lost in the faeces (which is said to contain 10^{11} bacteria per gram). The **bacteria** are involved in the **synthesis** of **vitamins K, B$_{12}$, thiamine** and **riboflavin**, the **breakdown** of **primary** to **secondary bile acids** and the **conversion** of **bilirubin** to **non-pigmented metabolites**, all of which are readily absorbed by the GI tract. The bacteria also break down cholesterol, some food additives and drugs.

38 Endocrine control

Multicellular organisms must coordinate the diverse activities of their cells, often over large distances. In animals, such coordination is achieved by the nervous and **endocrine** systems, the former providing rapid, precise but short-term control and the latter providing generally slower and more sustained signals. The two systems are intimately integrated and in some places difficult to differentiate. Endocrine control is mediated by **hormones**, signal molecules usually secreted in low concentrations (10^{-12}–10^{-7} M) from ductless glands into the bloodstream, so that they can reach all parts of the body. Other types of hormonal communication are mediated over smaller distances. **Paracrine** cells release their hormones into the immediate environment of the secretory cell, **autocrine** cells release hormones to control their own function, and **juxtacrine** communication requires direct physical contact between signal chemicals on the surface of one cell and receptor molecules on the surface of a neighbour. Some hormones are not secreted by discrete glands (Table 38). For instance, several of the cytokines released by immune cells (Chapter 10) act at some distance from their site of release and can fairly be considered as hormones.

Features of hormonal signalling

Hormonal molecules can be: (i) **modified amino acids** (e.g. adrenaline (norepinephrine); Chapter 45); (ii) **peptides** (e.g. somatostatin; Chapter 40); (iii) **proteins** (e.g. insulin; Chapter 39); or (iv) derivatives of the fatty acid cholesterol, such as **steroids** (e.g. cortisol; Chapter 45; Table 38). Protein and peptide hormones are cleaved from larger gene products, whereas smaller molecules require the precursor to be transported into endocrine cells so that it can be modified by sequences of enzymes to generate the final product (e.g. Chapter 45). Most hormones are stored in intracellular membrane-bound **secretory granules**, to be released by a **calcium-dependent mechanism** similar to the release of neurotransmitters from nerve cells (Chapter 8; Fig. 39b) when the cell is activated. However thyroid hormones and steroids, which are highly lipid soluble, cannot be stored in this way. Most steroids are made immediately prior to release, whereas the thyroid hormones are bound within a glycoprotein matrix (Chapter 41). After secretion, many hormones bind to **plasma proteins**. In most cases this involves non-specific binding to albumin, but there are **specific binding proteins** for some hormones, such as cortisol or testosterone. A hormone bound to a plasma protein cannot reach its site of action and is protected from metabolic degradation, but is freed when the plasma level of the hormone falls. The bound fraction thus acts as a reservoir that helps to maintain steady plasma levels of the free hormone.

Hormones exert their effects by interactions with **specific receptor proteins** and will act only on cells carrying those receptors. Most hormones activate cell surface receptors that are coupled to guanosine triphosphate-binding proteins (**G-proteins**) (Chapter 5) or that have **intrinsic tyrosine kinase** activity (e.g. Chapter 39). Receptors for lipid-soluble hormones (steroids, thyroid hormones) are usually *inside* the target cell, and modify gene transcription directly (e.g. Chapter 41). Because they are in the bloodstream, free hormones can reach all of the tissues that bear the appropriate receptors. Endocrine signals therefore provide a good way of inducing simultaneous changes in multiple organs, and most hormones have effects in more than one tissue. A corollary of this position is that many physiological processes are influenced by more than one hormone, as will become clear in subsequent chapters. Hormones are **inactivated** by metabolic transformation by enzymes, usually in the liver or at the site of action. It is a general rule that the smaller the hormone, the more rapid its inactivation.

Control of hormones

Endocrine glands can be controlled by the nervous system, other endocrine glands, or respond directly to levels of metabolites in the environment of the gland, and most are subject to all of these types of control. A common feature of hormonal control systems is a heavy reliance on **negative feedback loops**. Almost all hormones feed back to inhibit their own release, providing a direct method of moderating the output of hormone into the blood (Chapters 40–49). A less common feature of endocrine systems, associated only with reproductive functions, is *positive* feedback, whereby the release of a hormone leads to events that further promote release (Chapters 46, 47 & 49). The relatively slow nature of hormonal signalling puts limits on the types of physiological processes that can be controlled by hormones. They fall into four broad categories: (i) **homeostasis**; (ii) **reproduction**; (iii) **growth and development**; and (iv) **metabolism**. These systems work over time-scales that range from a few minutes (e.g. milk ejection; Chapter 49) to years (growth; Chapter 43).

Table 38 Tissues involved in endocrine control systems. The top half of the table shows the products of classical glandular tissue; the bottom half lists some of the other organs that release hormones

Secreting tissue	Main hormone(s)	Main target tissue(s)	Discussed in Chapter
Glands			
Anterior pituitary (P)	Adrenocorticotrophic hormone (ACTH)	Adrenal cortex (M)	45
	Growth hormone (GH)	Liver, bones, muscle (G)	43
	Follicle-stimulating hormone (FSH)	Gonads (R)	
	Luteinizing hormone (LH)	Gonads (R)	46
	Prolactin	Mammary glands (R)	49
	Thyroid-stimulating hormone (TSH)	Thyroid gland (G, M)	41
Intermediate pituitary (P)	Melanotrophin-stimulating hormone (MSH)	Melanocytes (H)	40
Posterior pituitary (P)	Antidiuretic hormone (ADH)	Kidney (H)	31
	Oxytocin	Mammary glands	49
		Uterus (R)	48
Pineal (A)	Melatonin	Hypothalamus (H)	
Thyroid (A)	Thyroxine (T_4)	Most tissues (G, M)	42
	Tri-iodothyronine (T_3)	Most tissues (G, M)	
	Calcitonin	Bones, gut (H)	44
Parathyroid (P)	Parathyroid hormone (PTH)	Bones, gut (H)	44
Pancreas (P)	Insulin	Liver, muscle, adipose tissue (G, M, H)	39
	Glucagon		
Adrenal cortex (S)	Corticosteroids (including cortisol)	Multiple (G, M)	45
	Aldosterone	Kidney (H)	31 & 45
Adrenal medulla (A)	Adrenaline (epinephrine)	Multiple (H, M)	45
	Noradrenaline (norepinephrine)	Multiple (H, M)	
Gonads: male (S)	Testosterone	Testes (R)	46
Gonads: female (S)	Oestradiol	Ovaries, uterus (R)	46 & 48
	Progesterone	Ovaries, uterus (R)	
Placenta (P, S)	Human chorionic gonadotrophin (hCG)	Uterus (R)	48
	Oestradiol	Ovaries, uterus (R)	
	Progesterone	Ovaries, uterus (R)	
Non-glands			
Brain (P, A)	Hypothalamic-releasing hormones	Anterior pituitary gland (H, R, M)	40
	Growth factors	Various (M)	
Heart (P)	Atrial natriuretic peptide (ANP)	Kidney	31
Kidney (P, S)	Erythropoietin (EPO)	Bone marrow (M)	9
	1,25-Dihydroxycholecalciferol	Gut, kidney (H)	44
	Renin	Plasma proteins (H)	31
Liver (P)	Insulin-like growth factor-1 (IGF-1)	Various (M)	43
Adipose tissue (P)	Leptin	Hypothalamus (M)	40
Gastrointestinal tract (P, A)	Gastrin	Gut (H, M)	33–37
	Secretin		
	Cholecystokinin (CCK)		
	Vasoactive intestinal polypeptide (VIP)		
	Gastrin-releasing peptide (GRP)		
Immune cells (P)	Cytokines	Hypothalamus (H, M)	10
	Opioid peptides		
Platelets (P)	Growth factors	Various (G)	42
Various sites (P)	Growth factors	Various (G)	42
	Neurotrophins	Neurones (G)	

Molecules: A, modified amino acid; P, peptide/protein; S, steroids/sterols.
Functions: H, homeostasis; R, reproduction; G, growth and development; M, metabolism.

39 Control of metabolic fuels

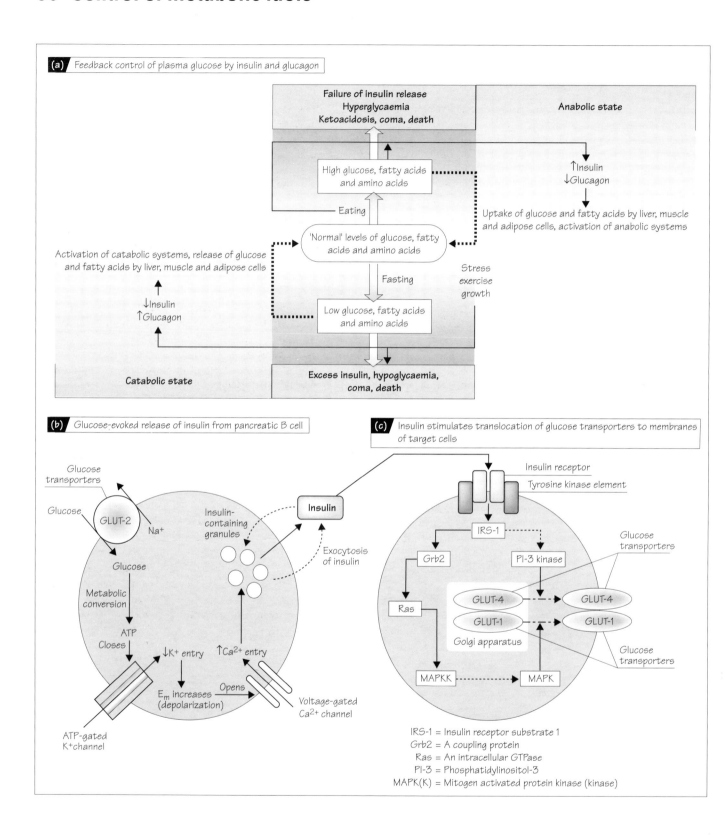

(a) Feedback control of plasma glucose by insulin and glucagon

Failure of insulin release
Hyperglycaemia
Ketoacidosis, coma, death

Anabolic state

High glucose, fatty acids
and amino acids

↑Insulin
↓Glucagon

Eating

Uptake of glucose and fatty acids by liver, muscle
and adipose cells, activation of anabolic systems

Activation of catabolic systems, release of glucose
and fatty acids by liver, muscle and adipose cells

'Normal' levels of glucose, fatty
acids and amino acids

Stress
exercise
growth

↓Insulin
↑Glucagon

Fasting

Low glucose, fatty acids
and amino acids

Catabolic state

Excess insulin, hypoglycaemia,
coma, death

(b) Glucose-evoked release of insulin from pancreatic B cell

Glucose
transporters

Glucose

GLUT-2

Na+

Insulin-
containing
granules

Insulin

Exocytosis
of insulin

Glucose

Metabolic
conversion

ATP
Closes

↓K+ entry

↑Ca2+ entry

E_m increases
(depolarization)

Opens

Voltage-gated
Ca2+ channel

ATP-gated
K+channel

(c) Insulin stimulates translocation of glucose transporters to membranes
of target cells

Insulin receptor

Tyrosine kinase element

IRS-1

Grb2

PI-3 kinase

Glucose
transporters

Ras

GLUT-4

GLUT-4

GLUT-1

GLUT-1

Golgi apparatus

Glucose
transporters

MAPKK

MAPK

IRS-1 = Insulin receptor substrate 1
Grb2 = A coupling protein
Ras = An intracellular GTPase
PI-3 = Phosphatidylinositol-3
MAPK(K) = Mitogen activated protein kinase (kinase)

Animal cells utilize glucose and fatty acids as fuels to generate the energy-rich molecule adenosine triphosphate (ATP) (Chapter 4). The blood levels of these molecules must be carefully controlled to ensure a steady supply of fuel to active tissues, a task that is complicated by the tendency of animals to eat discrete meals rather than continuously. Immediately after a meal, circulating levels of fuel molecules rise and any excess to immediate requirements is stored. This requires the transport of the molecules into cells (primarily **liver**, **skeletal muscle** and the fat-storing cells of **adipose tissues**) and the synthesis of storage molecules, such as **glycogen**, a polymer of glucose, **triglycerides** (fats) and, to a lesser extent, proteins. As time after a meal increases, the consumption of blood glucose and fatty acids necessitates the activation of tissue energy stores. Glycogen is broken down into glucose, triglycerides are converted into free fatty acids and ketone bodies and, if the fast is prolonged, proteins are catabolized to provide a supply of amino acids that can be converted to glucose (**gluconeogenesis**). The body thus alternates between two states, which can be described as **anabolic**, in which storage molecules are manufactured, and **catabolic** in which the same molecules are broken down (Fig. 39a). Switching between these states is controlled mainly by hormones, with the pancreatic proteins **insulin** and **glucagon** being the prime movers of the anabolic and catabolic processes, respectively. In addition, growth hormone (Chapter 43), cortisol, adrenaline (epinephrine) and noradrenaline (norepinephrine) (Chapter 45) can stimulate catabolic processes (Fig. 39a). **Leptin** is a hormone released from adipose cells that signals the level of body fat stores to the hypothalamus, but its role in metabolic control is as yet uncertain.

Insulin and glucagon

These hormones are made in the endocrine tissues of the pancreas, known as the **islets of Langerhans**. Three main types of cell have been identified within the islets: peripherally located **A** (also known as α) cells, which manufacture and secrete **glucagon**; centrally located **B** (or β) cells for the production and release of **insulin**; and **D** (δ) cells that synthesize and liberate **somatostatin**. The exact role of somatostatin has not been established, but it may be involved in controlling the release of the other two hormones. Insulin release is stimulated initially during eating by the parasympathetic nervous system and gut hormones, such as secretin (Chapter 35), but most output is driven by the rise in plasma glucose concentration that occurs after a meal (Fig. 39a,b). Circulating fatty acids, ketone bodies and amino acids augment the effect of glucose. The major action of insulin is to stimulate glucose uptake, with the subsequent manufacture of glycogen and triglycerides by adipose, muscle and liver cells. Its effects are mediated by a **receptor tyrosine kinase** (RTK; Fig. 39c; Chapter 43). The enzyme activates an intracellular pathway that results in the translocation of the glucose transporters GLUT-4 and GLUT-1 to the plasma membrane of the affected cell, to facilitate the entry of glucose (Fig. 39c). Insulin thus decreases plasma glucose. Insulin release is reduced as the blood glucose concentration falls, and is further inhibited by catecholamines (Chapter 45) acting at B-cell

α$_2$-adrenoceptors (Chapters 8 & 45). Glucagon release patterns tend to be the mirror image of those of insulin. Low blood glucose initiates glucagon release directly and also drives nervous and hormonal release of catecholamines, which activate β-adrenoceptors (Chapters 8 & 45) on A cells to augment glucagon release. Glucagon acts on guanosine triphosphate-binding protein (G-protein)-coupled receptors that stimulate the production of intracellular cyclic adenosine monophosphate (cAMP) (Chapter 5). In liver cells, this results in the inhibition of glycogen synthesis and the activation of glycogen breakdown systems. Similar effects are obtained in muscle cells to increase circulating levels of glucose. There are interactions between glucagon and insulin within the islets: insulin inhibits A-cell release of glucagon, but glucagon *stimulates* the release of insulin, an effect that ensures a basal level of insulin release irrespective of glucose levels. The two hormones operate as part of a classical negative feedback system (Fig. 39a; Chapter 2), in which the A and B cells act as combined sensors–comparators, and their hormones activate the effector tissues.

Diabetes mellitus

This disease is caused by failure of B-cell function, either by **autoimmune attack**, in which the immune system (Chapter 10) misidentifies the cells as non-self and destroys them, or by pathologies, such as **obesity**, that impair insulin release. The former type of disease is usually early onset and is treated with insulin (**insulin-dependent diabetes**), whereas the latter develops later and is treated by drugs that stimulate insulin release (**non-insulin-dependent diabetes**). Untreated, the condition leads to chronically high levels of plasma glucose (**hyperglycaemia**), an early symptom of which is overloading of the kidney glucose transporters (Chapter 29) so that sugar begins to appear in the urine. The osmotic effect of glucose leads to excess production of urine (**polyuria**) that tastes sweet (this used to be the diagnostic test for diabetes, and gives the disease its name; Latin *mellitus* = sweet). Long-term hyperglycaemia drives excessive lipolysis by liver cells, leading to a build-up of ketone bodies and the condition known as **ketoacidosis**. This disrupts brain function, causing coma and eventually death. A sharp fall in blood glucose (**hypoglycaemia**) caused by excessive insulin administration starves the brain of its main metabolic fuel and, by a sad irony, can also lead to coma and death (Fig. 39a). The main symptoms of hyper- and hypoglycaemia are shown in Table 39.

Table 39 The main symptoms of hyper- and hypoglycaemia

	Hyperglycaemia	Hypoglycaemia
	Polydypsia (excessive thirst)	Increased appetite
	Polyuria	Headache
	Fatigue	Weakness
	Slow healing of wounds	Loss of concentration
	Impaired vision	Blurred vision
Extreme	Ketone breath	Coma
	Coma	

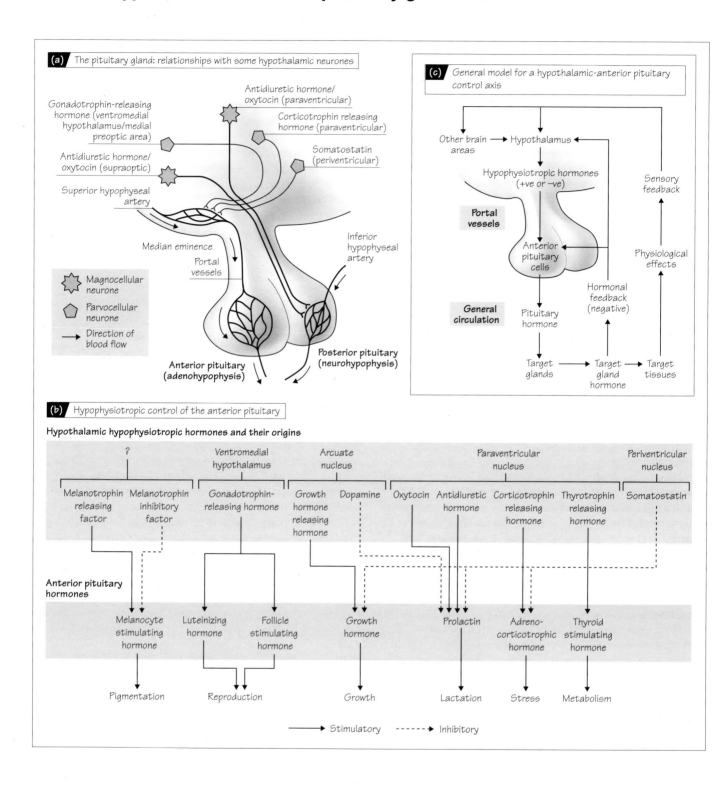

The **pituitary gland**, which is under the direct control of the brain from the **hypothalamus**, provides endocrine control of many major physiological functions. The hypothalamus is subdivided into a number of nuclei and vaguely defined 'areas' and surrounds the **third ventricle** at the base of the medial forebrain. The most important hypothalamic subdivisions for endocrine function are the **paraventricular**, **periventricular**, **supraoptic** and **arcuate nuclei**, and the **ventromedial hypothalamus**. The pituitary is located outside the skull immediately beneath the hypothalamus and comprises three divisions: the **anterior pituitary**, the **intermediate lobe** (almost vestigial in humans) and the **posterior pituitary** (Fig. 40a,b). The anterior glands develop from tissues originating in the roof of the mouth and are sometimes known as the **adenohypophysis**. The posterior gland grows downwards from the hypothalamus itself and is referred to as the **neurohypophysis**. All pituitary hormones are either peptides or proteins. As befits their developmental origins, the adeno- and neurohypophyses are controlled in different ways. However, a common feature of their regulation is that the hypothalamic neurones that control pituitary hormones are activated by neuronal signals and release chemicals in exactly the same way as other nerve cells (Chapter 8), albeit that their signals are liberated into the bloodstream rather than into synapses (Fig. 40a).

The anterior pituitary and intermediate lobe

The adenohypophyseal hormones and their actions are listed in Fig. 40b. They are released under the control of chemical signals (**hypophysiotropic hormones**) originating from small (**parvocellular**) neurones with their cell bodies in the hypothalamus (Fig. 40a–c). Hypophysiotropic hormones are peptides or proteins released into the blood at the **median eminence** (Fig. 40a) when the appropriate parvocellular neurones are electrically active. The blood supply for the anterior pituitary comes from the **superior hypophyseal artery**, in the form of the **hypophyseal portal vessels** (Fig. 40a). The portal vessels carry hypophysiotropic signals directly to the anterior pituitary to stimulate *or* inhibit the release of pituitary hormones by the activation of receptors on specific groups of pituitary cells (Fig. 40b). It should be noted that some hypophysiotropic hormones control more than one pituitary hormone. Figure 40c illustrates the basic principles that underlie the control of anterior pituitary hormones; this is a form of chemical cascade that allows for the precise control of pituitary output with two stages of signal amplification: first, at the pituitary itself, where tiny amounts of hypophysiotropic hormones control the release of larger quantities of pituitary hormone; and then at the final target gland, where the pituitary signals stimulate the release of still larger quantities of hormones such as steroids. The cascade allows for feedback control of hormone release at several points. The final hormone (and often some of the intermediate signals) inhibits further activity in the axis to provide the fine regulation of hormone release (Fig. 40c). This is a characteristic feature of anterior pituitary control systems.

The posterior pituitary

The posterior gland secretes two peptide hormones: **oxytocin** and **antidiuretic hormone** (**ADH**; also known as **vasopressin**). The hormones are manufactured in the cell bodies of large (**magnocellular**) neurones in the supraoptic and paraventricular nuclei of the hypothalamus, and are transported down the axons of these cells to their terminals on capillaries originating from the **inferior hypophyseal artery** within the posterior pituitary gland (Fig. 40a). When magnocellular neurones are activated (see Chapters 31, 48 & 49), they release oxytocin or ADH into the general circulation, from whence they can reach the relevant target tissues to produce the required effect. The signals that drive the release of posterior gland hormones are entirely neural, so that the hormones are said to be involved in **neuroendocrine reflexes**. These hormones operate over shorter time courses (minutes) than most endocrine events (hours to days). The release of ADH is controlled by conventional negative feedback mechanisms based on plasma osmolality (Chapter 31). Oxytocin, however, is involved in *positive* feedback mechanisms (Chapters 48 & 49).

Pulsatile release of pituitary hormones

Hormones released from the hypothalamus tend to appear in the blood in discrete pulses, rather than as continuous secretions. This is achieved by the synchronous activation of hormone-releasing neurones of the hypothalamus. As will be seen in later chapters, episodic release has profound implications for the operation of the endocrine system. It also raises a number of interesting and as yet unanswered questions as to how many separate and more or less widely scattered neurones can be activated simultaneously to give rise to pulsatile release.

41 Thyroid hormones and metabolic rate

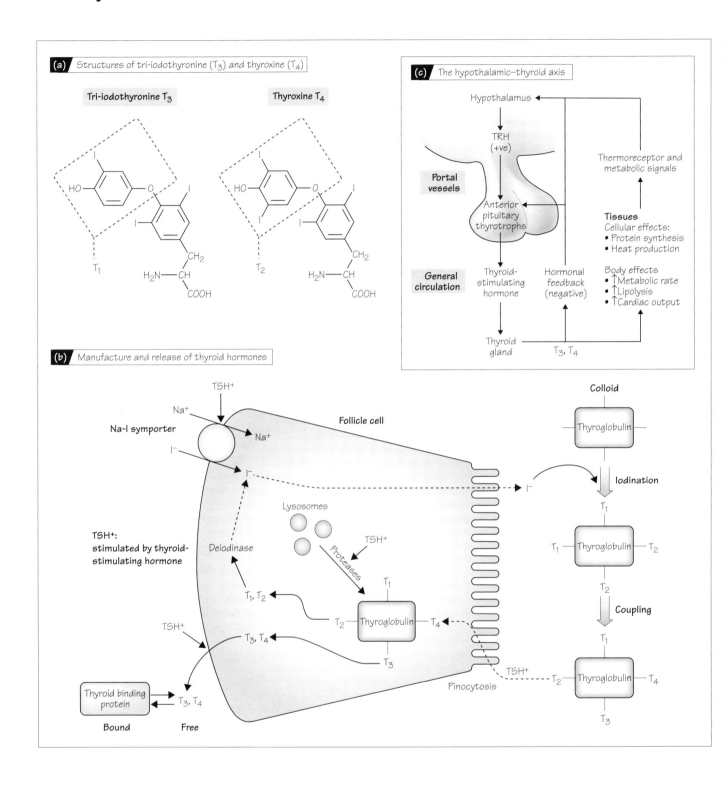

(a) Structures of tri-iodothyronine (T₃) and thyroxine (T₄)

(b) Manufacture and release of thyroid hormones

(c) The hypothalamic–thyroid axis

The thyroid gland is attached to the anterior surface of the trachea just below the larynx. It releases two iodine-containing hormones, **thyroxine** (also known as T_4) and **tri-iodothyronine** (T_3; Fig. 41a), the main effect of which is to increase heat production (**thermogenesis**) throughout the body and thereby induce an increase in metabolic rate. The hormones also have a crucial role in growth and development.

Synthesis and release

The thyroid gland is formed from clusters of cells (**follicles**) that surround a gel-like matrix or colloid, the primary constituent of which is the glycoprotein **thyroglobulin**. The follicle cells actively accumulate iodide (I^-) ions by means of an Na^+–I^- symporter (Chapter 5) driven by the inward sodium gradient (Fig. 41b). The formation of T_3 and T_4 occurs in two steps: (i) the amino acid tyrosine is iodinated to form mono- (T_1) or di-iodotyrosine (T_2) (Fig. 41a); (ii) T_2 is then coupled to T_1 or T_2 by thyroperoxidase to form the thyroid hormones. This process occurs with the tyrosine residues attached to thyroglobulin, so that, at any one time, this protein is festooned with molecules of T_1, T_2, T_3 and T_4 (Fig. 41b). The thyroid hormones and their intermediates are highly lipophilic and would escape from the gland were they not bound to thyroglobulin, which thus acts as a nucleus for the manufacture of the hormones and as a storage site. The hormones are released under pituitary control via **thyroid-stimulating hormone (TSH)**, which is obligatory for normal thyroid function (Chapter 40; Fig. 41c). Under the action of TSH, thyroid follicle cells pinch off small quantities of colloid by **pinocytosis**. Lysozymal protease enzymes then act on the thyroglobulin to liberate the iodinated compounds into the cell and thence into the bloodstream (Fig. 41b). Free T_1 and T_2 are de-iodinated by enzymatic action before they can leave the cell. The average plasma concentration of T_3 is roughly one-sixth of that of T_4, and much of that derives from de-iodinated T_4. Most of the thyroid hormones in the blood are bound to thyroxine-binding protein and are thus unavailable to their receptors, which are located *inside* target cells, attached directly to deoxyribonucleic acid (DNA). The small amounts of free T_3 and T_4 in plasma readily cross the cell membranes to bind to thyroid hormone receptors (the most important of which is **$TR\alpha_1$**). Thyroid receptors are linked to a DNA sequence known as the **thyroid-response element (TRE)** which initiates the transcription of thyroid-responsive genes. T_3 is some 10 times more potent than T_4 in activating $TR\alpha_1$ and consequently mediates most thyroid hormone actions, notwithstanding its lower levels in plasma. Thyroid receptors are present in almost all tissues, with particularly high levels in the liver and low levels in the spleen and testes.

Physiological roles of thyroid hormones

Basal levels of thyroid hormone release are essential to maintain a normal metabolic rate. Situations requiring increased heat production, for instance when the core temperature falls, lead to enhanced activation of the thyroid axis. The effects take up to 4 days to reach a maximum, a slow time course that is characteristic of hormones acting through nuclear receptors. The primary action of thyroid hormones is an increase in the synthesis of sodium–potassium ATPase (Chapter 5), an enzyme that consumes large amounts of metabolic energy, to increase heat production. The hormones may also enhance the production of **uncoupling proteins (UCPs)**. These molecules act in mitochondria to divert the H^+ ion gradient generated by the electron transport chain (Chapter 5), so that it produces heat rather than driving adenosine triphosphate (ATP) synthase. Although UCP-1 is found only in brown fat, a tissue that is uncommon in adult humans, two other members of the family (UCP-2 and UCP-3) are present in muscle and other tissues, and may thus contribute to thyroid-stimulated thermogenesis. Other important actions of thyroid hormones include a generalized increase in protein turnover (i.e. breakdown *and* synthesis), an increase in cardiac output caused by the enhancement of the effects of adrenaline (epinephrine) at β-adrenoceptors (Chapters 8 & 45), and a strong lipolytic effect that arises from the potentiation of responses to cortisol, glucagon, growth hormone and adrenaline. These actions can be described as generally catabolic (Chapter 39), but it should be noted that low doses of thyroid hormones have an overall anabolic action and that the hormones are essential to normal postnatal growth.

Disorders of the thyroid gland

Lack of dietary iodide or a failure of iodide uptake mechanisms in the thyroid gland produces the conditions of **hypothyroidism**. In juveniles, underproduction of thyroid hormones causes inadequate somatic and neural development and gives rise to **cretinism**, a condition characterized by subnormal stature and mental ability. In adults, the main symptoms of thyroid insufficiency are lethargy, sluggishness and an intolerance to cold. In severe cases, there is excess production of water-retaining mucoproteins in subcutaneous tissues, giving rise to tissue bloating, known as **myxoedema**. Such conditions are treated with injections of T_4. When the cause of hypothyroidism is an insufficiency of iodide intake, cells of the thyroid gland undergo hypertrophy and the gland becomes enlarged to form a **goitre**. This (now very uncommon) condition is treated by ensuring an adequate supply of dietary iodide. The overproduction of T_3 and T_4 leads to **hyperthyroidism**, characterized by **exophthalmia** (bulging eyes), increased behavioural excitability, tremor, weight loss and chronic tachycardia (high heart rate). The last of these symptoms can eventually lead to ventricular arrhythmias and/or heart failure, and so treatment, usually surgical removal of part of the gland or antithyroid drugs, is highly recommended.

42 Growth factors

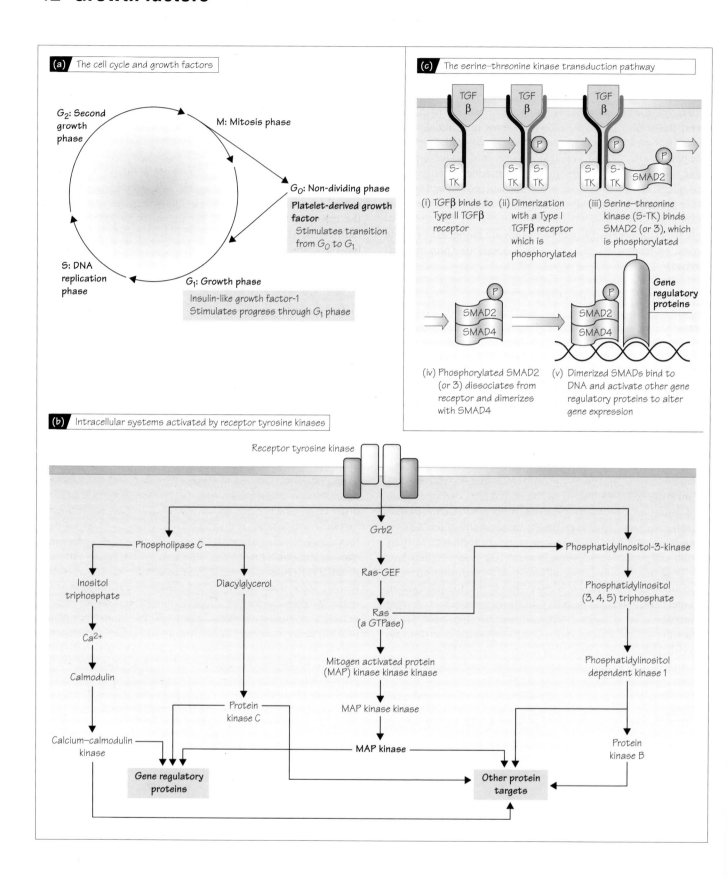

(a) The cell cycle and growth factors

G_2: Second growth phase

M: Mitosis phase

G_0: Non-dividing phase

Platelet-derived growth factor
Stimulates transition from G_0 to G_1

S: DNA replication phase

G_1: Growth phase

Insulin-like growth factor-1
Stimulates progress through G_1 phase

(c) The serine–threonine kinase transduction pathway

TGFβ   TGFβ   TGFβ

S-TK   S-TK S-TK   S-TK S-TK   SMAD2

(i) TGFβ binds to Type II TGFβ receptor

(ii) Dimerization with a Type I TGFβ receptor which is phosphorylated

(iii) Serine–threonine kinase (S-TK) binds SMAD2 (or 3), which is phosphorylated

SMAD2   SMAD2
SMAD4   SMAD4

Gene regulatory proteins

(iv) Phosphorylated SMAD2 (or 3) dissociates from receptor and dimerizes with SMAD4

(v) Dimerized SMADs bind to DNA and activate other gene regulatory proteins to alter gene expression

(b) Intracellular systems activated by receptor tyrosine kinases

Receptor tyrosine kinase

Grb2

Phospholipase C

Phosphatidylinositol-3-kinase

Inositol triphosphate

Diacylglycerol

Ras-GEF

Phosphatidylinositol (3, 4, 5) triphosphate

Ca^{2+}

Ras (a GTPase)

Calmodulin

Mitogen activated protein (MAP) kinase kinase kinase

Phosphatidylinositol dependent kinase 1

Protein kinase C

MAP kinase kinase

Calcium–calmodulin kinase

MAP kinase

Protein kinase B

Gene regulatory proteins

Other protein targets

For an embryo to develop into an adult, its cells must increase in number by the process of division (**mitosis**) and grow in size (**hypertrophy**). As they mature, cells develop specializations according to the tissue of which they are a part (**differentiation**). In tissue development, excess production of cells is the norm, so that the final shaping of organs depends on programmed death (**apoptosis**) of supernumerary cells. Some tissues, such as nerve and skeletal muscle, reach a stage of terminal differentiation in adulthood and undergo no further cell division. However, most cells in the adult retain the ability to divide, allowing tissues (e.g. blood vessels, bones) to remodel or repair themselves as required. Some cells face particularly high rates of attrition (e.g. enterocytes in the gut lining, skin cells, hair follicles) and are produced continuously throughout life. The processes of mitosis, cell growth and apoptosis are controlled by a large number of systemic and local peptide hormones, known as **growth factors** (Chapter 38). To varying extents, these factors stimulate mitosis (they are **mitogens**), promote growth (a **trophic** effect) and inhibit apoptosis (promote cell **survival**).

Growth factor families and their receptors

Growth factors are classified into a number of families based on common amino acid sequences and the types of receptor that they activate. **Neurotrophins**, which include **nerve growth factor** (**NGF**), are important chemical signals in the development of the nervous system and are potent survival factors for neurones in adults. The **epidermal growth factor** (**EGF**) family includes EGF itself and transforming growth factor-α (TGFα), both of which are mitogens in a wide range of tissues, including the gut and skin. **Fibroblast growth factors** (**FGF-1–24**) are strongly mitogenic and induce the production of new blood vessels (**angiogenesis**). The **transforming growth factor-β** (**TGFβ**) superfamily includes a number of bone transforming proteins (Chapter 43) and is crucial in embryogenesis and the development and remodelling of structural tissues. The origins of **platelet-derived growth factor** (**PDGF**) are self-explanatory. It stimulates division, growth and survival in a number of cell types and is important in tissue repair after injury. **Insulin** and **insulin-like growth factors** (**IGF-1** and **IGF-2**) have similar structures but rather different actions: insulin promotes anabolic activity generally (Chapter 40), whereas the IGFs are mitogenic, trophic and act as survival factors for several cell types. Numerous other hormones have mitogenic properties: for instance, the stimulation of red blood cell production by **erythropoietin** (Chapter 9) and white cell production by **cytokines** (Chapter 10) means that these hormones are also described as growth factors.

Mitosis occurs during the **cell cycle** (Fig. 42a). Some mitogens, including PDGF, stimulate transition from the non-dividing state (G_0) into the growth phase of the cycle (G_1), whereas others, such as EGF and IGF-1, stimulate progress through G_1. With the exception of TGFα, erythropoietin and the cytokines, growth factors work by activating receptor tyrosine kinases (Chapter 39; Fig. 42b). Binding of the hormone leads to phosphorylation of the tyrosine residues of a number of important intracellular proteins, including phospholipase C, Grb2 and phosphatidylinositol-3 kinase, eventually leading to the production of more kinases: **protein kinases C and B, calcium-calmodulin kinase** (**CAM-kinase**) and **mitogen-activated protein kinase** (**MAP-kinase**) (Fig. 42b). These enzymes have many targets within the cell, but MAP-kinase, in particular, enters the nucleus and activates immediate to early genes, such as *c-fos* and *c-jun*. The products of these genes are transcription factors (Chapter 1), driving the expression of further genes, such as those that produce **G_1 cyclins**, proteins that are required for cell division. The MAP-kinase pathway appears to be the main intracellular signalling system for the stimulation of mitosis. The TGFβ family exerts its effects through **receptor serine–threonine kinases** that phosphorylate their target proteins at serine and threonine residues. The pathway activated by these receptors involves proteins called **SMADs** (the name is derived from genes that code for similar proteins in *Drosophila melanogaster* (fruit fly) and *Caenorhabditis elegans*, a nematode worm). SMAD-2 and/or SMAD-3 is phosphorylated whilst it is attached to the receptor; it then dissociates to dimerize with SMAD-4, forming a complex that directly activates gene regulatory proteins (Fig. 42c). Growth hormone, erythropoietin and the cytokines activate receptors that signal through **Janus kinases** (**JAKs**; see Chapter 44).

Growth factors and cancer

Cell division and growth are strictly controlled so that organs do not invade the space needed for other tissues. When this process is deranged, cancers are formed. Cancer cells do not recognize the normal constraints of organ growth or the limits to the number of divisions to which cells are normally subjected, and are unusually mobile. These features make cancer cells extremely dangerous, as they supplant healthy tissues and cause fatal damage to physiological systems. Cancerous growths start with mutations in particular genes (**oncogenes**) that impact on cell division and/or apoptosis. *Ras* genes, which produce the Ras GTPases that are key mediators in the MAP-kinase pathway (Fig. 42b), are commonly found to be defective in human tumours. In view of the importance of this pathway in mitogenesis, it is not difficult to see how the abnormal activation of these genes could lead to excessive cellular proliferation. In this situation, the signals involved in normal tissue growth provide the driving force for tumour growth and survival. EGF, in particular, has been associated with the maintenance of colorectal and breast cancers, and anti-EGF drugs are showing some promise as tumour-controlling agents.

43 Somatic and skeletal growth

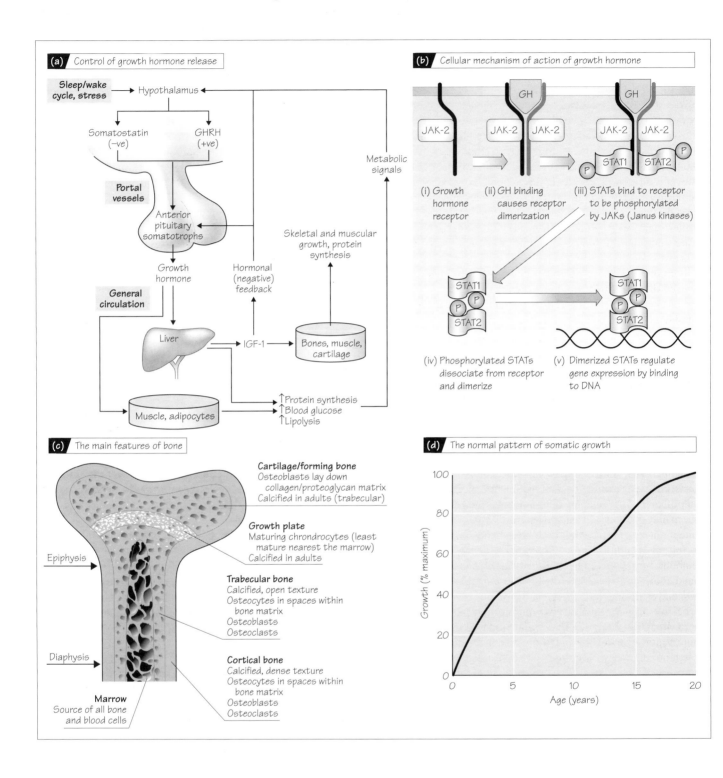

(a) Control of growth hormone release

Sleep/wake cycle, stress → Hypothalamus

Somatostatin (−ve)   GHRH (+ve)

Portal vessels

Anterior pituitary somatotrophs

Growth hormone

Hormonal (negative) feedback

General circulation

Liver → IGF-1 → Bones, muscle, cartilage

Skeletal and muscular growth, protein synthesis

Metabolic signals

Muscle, adipocytes → ↑Protein synthesis
↑Blood glucose
↑Lipolysis

(b) Cellular mechanism of action of growth hormone

GH

JAK-2   JAK-2 JAK-2   JAK-2 JAK-2

P STAT1   STAT2 P

(i) Growth hormone receptor

(ii) GH binding causes receptor dimerization

(iii) STATs bind to receptor to be phosphorylated by JAKs (Janus kinases)

STAT1 P P STAT2

STAT1 P P STAT2

(iv) Phosphorylated STATs dissociate from receptor and dimerize

(v) Dimerized STATs regulate gene expression by binding to DNA

(c) The main features of bone

Epiphysis

Diaphysis

Marrow
Source of all bone and blood cells

Cartilage/forming bone
Osteoblasts lay down collagen/proteoglycan matrix
Calcified in adults (trabecular)

Growth plate
Maturing chondrocytes (least mature nearest the marrow)
Calcified in adults

Trabecular bone
Calcified, open texture
Osteocytes in spaces within bone matrix
Osteoblasts
Osteoclasts

Cortical bone
Calcified, dense texture
Osteocytes in spaces within bone matrix
Osteoblasts
Osteoclasts

(d) The normal pattern of somatic growth

Growth (% maximum) vs Age (years), axis 0–100, 0–20

Growth and development are entirely under endocrine control. The key signals involved in these processes are **growth hormone**, the thyroid hormones (Chapter 41), sex steroids (Chapters 46 & 47) and the growth factors (Chapter 42). Normal growth depends on the interplay between all of these factors. In development, there are two periods of particularly rapid growth: the 2 years immediately after birth, and at the age of puberty (Fig. 43d).

Growth hormone

Growth hormone (also known as **somatotrophin**) provides the main drive for the growth spurts in development. It is a protein released from pituitary somatotrophs under hypothalamic control (Fig. 43a) that stimulates growth in muscles, bones and connective tissue. It is essential for normal growth both pre- and postnatally. The release of the hormone increases immediately after birth before subsiding to a low level for most of prepubertal life. There is another surge in release around the time of puberty, after which plasma concentrations again fall to remain steady for the next 50 years or so. After this time, levels decline gradually. The release of the hormone varies throughout the day, with the highest levels achieved during deep sleep. The episodic appearance of growth hormone in the blood is driven by hypothalamic **growth hormone-releasing hormone** (**GHRH**), which peaks at night, and **somatostatin** (**SST**), which appears in portal blood mainly during the day and inhibits growth hormone release (Chapter 40; Fig. 43a). The growth hormone receptor is linked to an intracellular enzyme, **Janus kinase-2** (**JAK-2**) (Fig. 43b). Once activated, this enzyme binds and phosphorylates **signal transduction and activation of transcription** (**STAT**) proteins, which consequently modify gene transcription. To provide energy for growing tissues, growth hormone has an anti-insulin action in increasing plasma glucose and stimulating lipolysis (Chapter 39). However, its overall effect is anabolic, increasing protein synthesis in many tissues. Most of its effects on growth arise from the stimulation of the release of insulin-like growth factor-1 (IGF-1) (Chapter 42) into the circulation, mainly from the liver. The lifetime release of growth hormone is regulated by the genetic factors that determine body size, but full expression of its effects requires adequate supplies of metabolic fuels and the presence of the other hormones mentioned above. In the short term, it is also liberated in response to stress and exercise.

The overproduction of growth hormone in children is associated with **gigantism**, and underproduction with **dwarfism**, which is much more common. Dwarfism is currently treated by human growth hormone manufactured by genetically engineered bacteria. Excess growth hormone release in adults leads to disproportionate growth of the bones of the face and limb extremities, a condition known as **acromegaly**.

Bone growth and remodelling

The bones are a major target for growth hormone. They are composed of an organic matrix made up of the structural protein **collagen**, combined with glycoproteins, that forms a framework within which the mineral **hydroxyapatite** ($Ca_{10}(PO_4)_6(OH)_2$) is deposited. There are two main varieties of bone structure. **Cortical** or compact bone has a dense structure and provides most of the strength of the skeleton. It forms the outer layer of all bones and is particularly prevalent in the **diaphyses** (shafts) of limb bones. **Trabecular** or spongy bone has a more open structure than cortical bone and surrounds the marrow. Axial bones, such as the vertebrae, and the ends (**epiphyses**) of long bones are largely composed of trabecular matrix (Fig. 43c). In development, bones grow from the interface between the epiphysis and the diaphysis (the **growth plate**). The elongation of bones involves the laying down of new collagen matrix at the growth plate by rapidly dividing **chondrocytes**, followed by **calcification** (hydroxyapatite deposition) through the action of **osteoblasts**. When growth is complete at about 20 years of age (Fig. 43d), the growth plate itself becomes calcified and bone elongation ceases. This stage is known as **epiphyseal closure**, a process driven by the high levels of sex steroids present at puberty. Even in adults, bones remain dynamic structures, with substantial proportions of the skeleton (25% of trabecular bone and 3% of cortical bone) being replaced by new growth every year. Osteoblasts develop into **osteocytes**, cells with numerous processes that settle into spaces in the bone matrix. Osteocytes maintain the integrity of the matrix, but can also secrete acids that dissolve hydroxyapatite and thus provide free Ca^{2+} to the circulation when required (Chapter 44). **Osteoclasts** are large cells similar to macrophages (Chapter 10) that remove old bone matrix so that it can be replaced by new material. Osteoblasts, osteocytes and osteoclasts are all present in mature bone. The collective activity of these cells allows bone to be remodelled throughout life to cope with changes in skeletal stresses, and plays an essential role in the repair of broken bones. All bone cells differentiate from bone marrow stem cells. Systemic IGF-1 and locally produced IGF-1 and IGF-2 (Chapter 42) stimulate the division, differentiation and matrix-secreting activity of osteoblasts and chondrocytes (which are also involved in cartilage formation), whereas members of the transforming growth factor-β (TGFβ) family of growth factors are thought to provide the same stimuli for osteoclasts.

Osteoporosis

Following the menopause, women lose bone mass, leading to a weakening of the skeleton with a consequent increase in the likelihood of fractures in older women. This is due to the reduced secretion of sex steroids from the ovaries (Chapter 46), which normally suppress the production of the cytokine **interleukin-6 (IL-6)** in bones. High levels of IL-6 stimulate the differentiation of osteoclasts, so that bone resorption outstrips the laying down of new matrix and more bone is removed than is replaced. The condition can be successfully treated by the administration of oestrogen (**hormone replacement therapy**). Recent evidence suggests that bone destruction in **rheumatoid arthritis** may also be driven by cytokines.

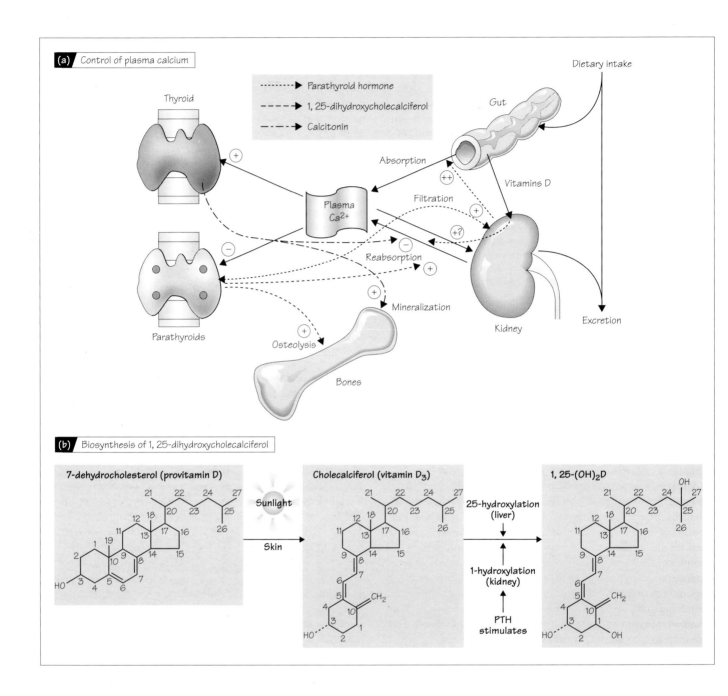

(a) Control of plasma calcium

(b) Biosynthesis of 1, 25-dihydroxycholecalciferol

In cells, calcium (Ca^{2+}) ions are used to trigger many key physiological events, including muscle contraction (Chapter 50), the release of neurotransmitters (Chapter 8), the release of hormones (Chapters 39 & 40), secretion from exocrine glands and the activation of many important intracellular enzymes, such as the calcium-calmodulin kinases (CAM-kinases) (Chapter 42) and nitric oxide synthase (Chapters 17 & 47). Ca^{2+} ions are triggers in so many crucial events that free intracellular Ca^{2+} must be maintained at a very low level (Chapter 3). Most is stored in the endoplasmic reticulum or mitochondria. External to cells, Ca^{2+} ions contribute to the blood clotting cascade (Chapter 9) and the normal functioning of Na^+ ion channels (Chapter 5). When extracellular Ca^{2+} is too low, Na^+ channels open spontaneously, leading to involuntary contractions of skeletal muscles, described as **hypocalcaemic tetany**. This is the clinical sign of low plasma Ca^{2+}. It is evident that Ca^{2+} levels in plasma must be very carefully controlled, a function performed by the coordinated activity of three hormones: **parathyroid hormone (PTH)**, **1,25-dihydroxycholecalciferol (1,25-($OH)_2$D)** and **calcitonin** (Fig. 44a).

Parathyroid hormone and calcitonin

PTH, an 84-amino-acid peptide, is the major controller of free calcium in the body. It is released from **chief cells** of the four (or more) **parathyroid glands** located immediately behind the thyroid gland, when the plasma concentration of Ca^{2+} decreases. The ion is detected by a membrane-bound receptor protein expressed by chief cells. When Ca^{2+} ions bind to the receptor, intracellular levels of cyclic adenosine monophosphate (cAMP) (Chapter 4) decrease and the release of PTH is *inhibited*. PTH increases the plasma levels of Ca^{2+} by activating specific membrane receptors in bone, gut and kidney. In bone, the immediate effect of PTH is to stimulate **osteocytic osteolysis** of bone crystals to release Ca^{2+} ions (Chapter 43). After a longer time, PTH also increases osteoclast activity (Chapter 43) to gain access to more of the bone mineral. In the gut, PTH, acting in concert with 1,25-dihydrocholecalciferol, enhances the absorption of Ca^{2+} ions. In the kidney, the same combination of hormones enhances the reabsorption of Ca^{2+} from the renal tubules and simultaneously decreases the reabsorption of phosphate ions (Chapter 30). PTH also stimulates the kidney to produce more 1,25-dihydrocholecalciferol. Thus, PTH leads the response to a fall in plasma Ca^{2+} by releasing ions stored in bone, conserving ions filtered by the kidney, and enhancing the intake of new ions from the gut (Fig. 44a). The effects of PTH are in each case mediated by stimulating an increase in cAMP in the target cells. **Calcitonin** is a 32-amino-acid peptide released from C cells of the thyroid gland in response to high levels of plasma Ca^{2+} ions. C cells carry the same Ca^{2+} receptor as parathyroid chief cells. Calcitonin inhibits bone resorption by osteocytes and may inhibit reabsorption in the kidney, so reducing plasma levels of the ion. The fact that complete removal of the thyroid gland causes no obvious problems with calcium homeostasis has led some physiologists to doubt the significance of this hormone in the normal control of Ca^{2+} ions.

Vitamin D and 1,25-dihydroxycholecalciferol

Vitamin D is an umbrella term for two molecules: **ergocalciferol** (vitamin D_2) and **cholecalciferol** (vitamin D_3). Both are derivatives of provitamin D (dehydrocholesterol; Fig. 44b) and the primary source of supply is the diet, with ergocalciferol derived from plants and yeasts and cholecalciferol from animal (particularly dairy) products. Unusually for a vitamin, cholecalciferol can be manufactured within the body via a reaction that is enabled by ultraviolet irradiation of the skin. The D vitamins are then converted to 1,25-dihydrocholecalciferol in the kidney (Fig. 44.1b). The final reaction is the slowest step in the process and therefore regulates the speed of the entire chain of reactions (i.e. it is rate limiting); it is under the influence of PTH. 1,25-dihydrocholecalciferol has a steroid-like structure and is sometimes referred to as a **sterol**. Its receptors are members of the superfamily of steroid receptors and are located *inside* target cells. The hormone–receptor complex binds to response elements on deoxyribonucleic acid (DNA) to drive the transcription of genes. **Calcium-binding protein**, which is thought to promote calcium transport across epithelia (Chapter 30), is the product of one of the genes activated by 1,25-dihydrocholecalciferol. The major action of 1,25-dihydrocholecalciferol is to enable Ca^{2+} absorption from the gut. Without the hormone, Ca^{2+} uptake is severely impaired to the point at which intake of the hormone is insufficient to maintain body stores. This leads to the increased release of PTH and resorption of bone. A lack of D vitamins in children leads to inadequate calcification of bones, which become malformed. This leads to the characteristically bowed limbs seen in **rickets**. This condition was common in the early part of the twentieth century, but was virtually eliminated in the UK by the introduction of free school milk. Insufficiency of vitamin D in adults leads to bone wasting, a condition known as **osteomalacia**, with symptoms similar to those of osteoporosis (Chapter 43). 1,25-dihydrocholecalciferol also promotes the reabsorption of Ca^{2+} from the kidney tubules. The effects of this hormone are generally augmented in the presence of PTH.

Other hormones affecting calcium

Growth-promoting hormones (growth hormone, thyroid hormones and sex steroids) tend to promote the incorporation of calcium into bones (see Chapter 43). Excess corticosteroids (Chapter 45) inhibit calcium uptake from the gut and reabsorption from the kidney.

45 The adrenal glands and stress

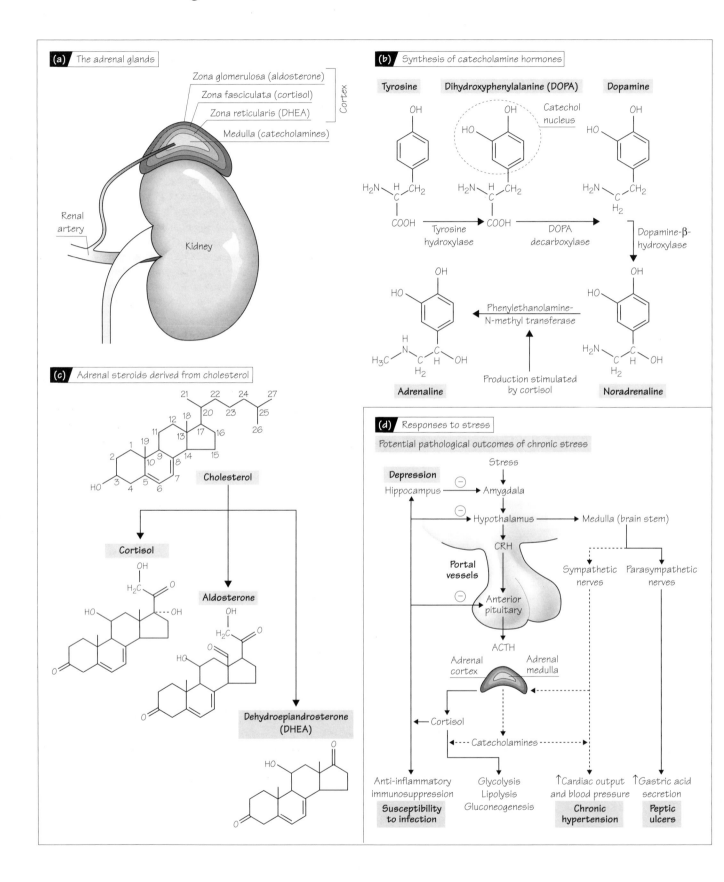

(a) The adrenal glands

Zona glomerulosa (aldosterone)
Zona fasciculata (cortisol)
Zona reticularis (DHEA)
Cortex
Medulla (catecholamines)

Renal artery

Kidney

(b) Synthesis of catecholamine hormones

Tyrosine Dihydroxyphenylalanine (DOPA) Dopamine

Catechol nucleus

Tyrosine hydroxylase → DOPA decarboxylase → Dopamine-β-hydroxylase

Phenylethanolamine-N-methyl transferase

Adrenaline

Production stimulated by cortisol

Noradrenaline

(c) Adrenal steroids derived from cholesterol

Cholesterol

Cortisol

Aldosterone

Dehydroepiandrosterone (DHEA)

(d) Responses to stress

Potential pathological outcomes of chronic stress

Stress

Depression
Hippocampus ⊖→ Amygdala
⊖→ Hypothalamus → Medulla (brain stem)
CRH
Portal vessels
⊖
Anterior pituitary
Sympathetic nerves   Parasympathetic nerves
ACTH

Adrenal cortex   Adrenal medulla

Cortisol

Catecholamines

Anti-inflammatory immunosuppression
Susceptibility to infection

Glycolysis
Lipolysis
Gluconeogenesis

↑Cardiac output and blood pressure
Chronic hypertension

↑Gastric acid secretion
Peptic ulcers

The adrenal glands are attached to each kidney (hence the name; Fig. 45a) and consist of two endocrine tissues of distinct developmental origins. The outer layers of the gland (the **adrenal cortex**) originate from mesodermal tissue and secrete steroid hormones, primarily under the control of the pituitary gland (Chapter 40). The inner core (the **adrenal medulla**) releases the catecholamine hormones **adrenaline** (epinephrine) and **noradrenaline** (norepinephrine). It develops from neuronal tissue and is functionally part of the **sympathetic nervous system** (Chapter 8). Removal of the adrenal glands in animals results in death within a few days, which is thought to result from the loss of the ability to cope with **stress**.

The adrenal medulla

The **chromaffin cells** of the adrenal medulla manufacture and secrete **noradrenaline** (20%) and **adrenaline** (80%). These catecholamine hormones are derived from tyrosine by a series of steps catalysed by specific enzymes (Fig. 45b). The production of the rate-limiting enzyme, **phenylethanolamine-N-methyl transferase**, is stimulated by **cortisol**, providing a direct link between the functioning of the medulla and cortex. The secretion of catecholamines is stimulated by sympathetic preganglionic neurones located in the spinal cord (Chapter 8), so that the adrenal medulla functions in concert with the sympathetic nervous system, of which noradrenaline is the main neurotransmitter. Catecholamine release contributes to normal physiological functions, but is enhanced by stress (see below). Adrenaline and noradrenaline act through guanosine triphosphate-binding protein (G-protein)-coupled **adrenoceptors**. These are classified as α_1, α_2 and β_1–β_3. The hormones have the same effects in tissues as in the stimulation of sympathetic nerves, with important stress-related responses being vasoconstriction (α_1), increased cardiac output (β_1) and increased glycolysis and lipolysis (β_2, β_3). These actions support increased physical activity. Noradrenaline has equal potency at all adrenoceptors, but adrenaline, at normal plasma concentrations, will only activate β-receptors (NB: higher levels do stimulate α-receptors). **Phaeochromocytoma** is a tumour of the adrenal medulla that leads to the excess production of catecholamines, with high blood pressure as the most immediately threatening symptom. It is treated by α-adrenoceptor antagonists and/or surgery.

The adrenal cortex

The cortex is made up of three zones of tissue: the outer **zona glomerulosa**, which releases **aldosterone**; the **zona fasciculata**, which produces **cortisol** and several related but less important hormones; and the inner **zona reticularis**, which secretes the androgen **dehydroepiandrosterone (DHEA)**. All of these secretions are steroids (Fig. 45c). Aldosterone is referred to as a **mineralocorticoid** as it controls the reabsorption of Na^+ and K^+ ions in the kidney (Chapter 31), whereas DHEA and its metabolite, **androstenedione**, provide an important source of androgens for females, contributing to hair growth and libido (Chapter 46). Cortisol and its analogues (such as cortisone) have powerful effects on glucose metabolism and are collectively classified as **glucocorti-** coids, although they do have some mineralocorticoid actions. The release of cortisol and DHEA is stimulated by **adrenocorticotrophic hormone (ACTH)** liberated from the pituitary gland (Chapter 40; Fig. 45d), whereas the secretion of aldosterone is stimulated by angiotensin II (Chapter 31). The effects of cortisol are mediated by intracellular receptors that translocate to the cell nucleus after binding the hormone. The cortisol–receptor complex binds to **glucocorticoid response elements** on deoxyribonucleic acid (DNA) to initiate gene transcription.

Cortisol is released during the course of normal physiological activity. The pattern of secretion is pulsatile, driven by activity in corticotrophin-releasing hormone neurones of the hypothalamus (Chapter 40). There is usually a surge in cortisol release in the hour after waking. The primary stimulus for the increased release of glucocorticoids is **stress**, which is the result of exposure to adverse situations. The **stress response** is driven by the **amygdala**, part of the forebrain that stimulates: (i) activity in hypothalamic CRH neurones; (ii) activity in the sympathetic nervous system; (iii) activity in the parasympathetic nerves that cause acid secretion in the stomach (Chapter 34); and (iv) the feeling of fear (Fig. 45d). The stress response evolved to cope with immediate threats, such as predators, to which the appropriate physiological reaction is to prepare for physical activity. The actions of the two parts of the adrenal gland are complementary in this respect. Catecholamines are released from the medulla to produce a rapid increase in cardiac output and the mobilization of metabolic fuels. Corticosteroids produce a slower, more sustained response, increasing the amount of glucose in the plasma (Chapter 39) by: (i) increasing glycolysis and gluconeogenesis in the liver (Chapter 36); (ii) reducing glucose transport into storage tissues; (iii) increasing protein catabolism with a consequent release of amino acids from all tissues other than the liver; and (iv) increasing the mobilization of lipids from adipose tissue. High levels of glucocorticoids also suppress the activity of immune cells to produce an anti-inflammatory effect, and can mimic the actions of aldosterone on the kidney to retain Na^+ and lose K^+ ions. The stress response is appropriate as long as the stress is relieved promptly. Unfortunately, modern life places many of us in positions in which stress is prolonged. This can lead to chronic hypertension, gastric ulceration, immunosuppression and depression (Fig. 45d). Glucocorticoid derivatives, such as **dexamethasone**, are widely used as anti-inflammatory agents in conditions such as arthritis and asthma. Chronically high levels of glucocorticoids eventually cause weakening of the skin, muscle wasting, reduction in bone strength, increased rates of infection due to immunosuppression, and can damage nerve cells in the **hippocampus** that are part of a feedback circuit controlling responses to stress (Fig. 45d). Thus, the long-term therapeutic use of steroids must be very carefully monitored, especially in the young where normal growth may be affected. Diseases of the adrenal cortex include **Cushing's syndrome**, which results from the excessive release of glucocorticoids and has a range of symptoms similar to those described above, and **Addison's disease**, which is the result of adrenocortical hypoactivity and is characterized by symptoms of hypoglycaemia, weight loss and skin pigmentation.

46 Endocrine control of reproduction

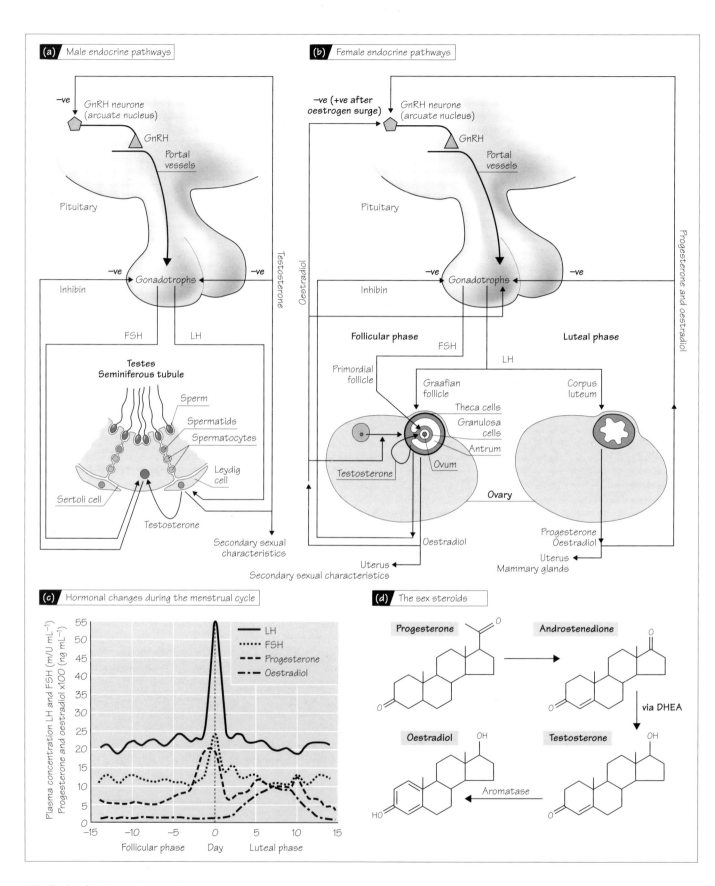

(a) Male endocrine pathways

−ve
GnRH neurone (arcuate nucleus)
GnRH
Portal vessels
Pituitary
−ve Gonadotrophs −ve
Inhibin
Testosterone
FSH   LH

Testes
Seminiferous tubule
Sperm
Spermatids
Spermatocytes
Leydig cell
Sertoli cell
Testosterone
Secondary sexual characteristics

(b) Female endocrine pathways

−ve (+ve after oestrogen surge)
GnRH neurone (arcuate nucleus)
GnRH
Portal vessels
Pituitary
−ve Gonadotrophs −ve
Inhibin
Oestradiol
Progesterone and oestradiol

Follicular phase   FSH   LH   **Luteal phase**

Primordial follicle
Graafian follicle
Theca cells
Granulosa cells
Antrum
Ovum
Testosterone
Corpus luteum
Ovary
Oestradiol
Progesterone Oestradiol
Uterus
Secondary sexual characteristics
Uterus
Mammary glands

(c) Hormonal changes during the menstrual cycle

Plasma concentration LH and FSH (m/U mL⁻¹)
Progesterone and oestradiol x100 (ng mL⁻¹)

LH
FSH
Progesterone
Oestradiol

55
50
45
40
35
30
25
20
15
10
5
0

−15 −10 −5 0 5 10 15
Follicular phase   Day   Luteal phase

(d) The sex steroids

Progesterone → **Androstenedione**
via DHEA
Oestradiol ← Aromatase **Testosterone**

Reproductive function in males and females is controlled by common hormonal systems based on the hypothalamic control of the pituitary **gonadotrophins**, individually known as **luteinizing hormone (LH)** and **follicle-stimulating hormone (FSH)**. These glycoproteins are released from the **gonadotrophs** of the anterior pituitary gland under the influence of **gonadotrophin-releasing hormone (GnRH**; Chapter 40) (Fig. 46). Failure of GnRH release is one cause of infertility. It is released in pulses at intervals of 1–3 h in males and females, a pattern that is accurately reflected in plasma levels of LH. The pulsatile pattern of GnRH secretion is essential for normal reproductive activity, as continuous exposure of gonadotrophs to the hormone leads to a rapid desensitization of these cells and a reduction in the release of gonadotrophins. The releasing hormone acts through receptors coupled to G_q (Chapter 4) to stimulate the release and manufacture of the gonadotrophins.

Actions of gonadotrophins

The gonadotrophins produce their effects via interactions with guanosine triphosphate-binding protein (G-protein)-coupled receptors that activate the intracellular production of cyclic adenosine monophosphate (cAMP) (Chapter 4). In the male, LH acts on the **Leydig cells** of the testes to stimulate the production of the steroid **testosterone**, which acts in concert with FSH on **Sertoli cells** of the **seminiferous tubules** to cause **spermatogenesis** (Fig. 46a). Sperm are generated in a two-stage meiosis from spermatocytes via spermatids. Spermatogenesis proceeds most efficiently at a temperature of 34°C, which is why the testes are located outside the body cavity. A normal adult male produces some 2×10^8 sperm per day, a process that carries on from puberty until the end of life. Sertoli cells also produce **inhibin**, a peptide feedback signal that specifically targets the release of FSH.

The situation in females varies over time according to the **menstrual cycle** (Fig. 46b,c), which lasts for around 28 days and is driven by central nervous pattern generators that set the activity of hypothalamic GnRH neurones. After puberty, the ovaries contain about 400 000 **primordial follicles**, each of which contains an **ovum** (or **oocyte**) in an arrested state of meiosis. All follicles are present at birth and no new gametes are formed after this time. Small groups of follicles begin to mature spontaneously throughout reproductive life, but only those for which development coincides with the appropriate phase of the cycle reach the stage of ovulation. In the first part of the cycle (the **follicular phase**), LH acts on **theca interna cells** in developing follicles to stimulate the production of testosterone, which is converted to **oestrogens** (mainly **oestradiol**; Fig. 46b) by **aromatase** enzymes in follicular **granulosa cells** under the influence of FSH. Granulosa cells also produce inhibin, which suppresses FSH release. In the follicular phase, oestrogens promote the growth of the uterine endometrial lining and the release of watery secretions at the cervix that enhance the transit of sperm into the uterus. Oestrogens also stimulate the production of LH receptors in granulosa cells. During this time, the

actions of FSH and oestrogens stimulate maturing follicles within the ovary, only the largest of which will normally undergo **ovulation**. The remainder wither away by the process of **atresia**. Ovulation occurs at about day 14 of the cycle (Fig. 46c). It is initiated by a large increase in the release of oestradiol from the granulosa cells, stimulated by their newly developed LH receptors. Normally, oestrogens act as a negative feedback signal, inhibiting LH release (Fig. 46b), but for 2 days after the oestrogen surge they *stimulate* LH release, i.e. the system switches from negative to *positive* feedback. This leads to a massive increase in the release of LH, which causes the wall of the most developed follicle to rupture and releases the ovum into the nearest **oviduct** to await fertilization (Chapter 48). Following ovulation, the theca cells of the ruptured follicle undergo **atrophy** (shrinkage) and the granulosa cells undergo **hypertrophy** (growth) to produce the **corpus luteum**, and the cycle enters the **luteal phase**. The corpus luteum produces **progesterone** (Fig. 46b), as well as oestrogens, in response to stimulation by LH. Progesterone prepares the reproductive tract for pregnancy, stimulating further growth of the uterine endometrium and altering the nature of cervical secretions to discourage the entry of sperm into the uterus. If fertilization does not occur, the corpus luteum undergoes **luteolysis** after roughly 14 days, a process that results from the reduced ability of LH to support the corpus, coupled with the increased secretion of **oxytocin** and **prostaglandins** from the uterus. In the absence of progesterone, leucocytes (Chapter 10) invade the endometrial lining which is shed in the process of **menstruation**, followed by the onset of a new cycle. After 30–40 years of menstrual activity, the exhaustion of ovarian follicles causes the female system to enter the **menopause**, after which reproduction is no longer possible. Circulating levels of sex steroids are greatly reduced, leading to drying of the secretory glands in the reproductive tract and other symptoms, including circulatory changes that cause hot flushes. The most pernicious outcome of the menopause is **osteoporosis** (Chapter 44).

All sex steroids exert their effects by interacting with intracellular receptors that bind to deoxyribonucleic acid (DNA) response elements, and thus induce changes in gene expression. Some of the actions of testosterone are actually mediated by **dihydrotestosterone**, produced within the target cells by the action of the enzyme **5-α-reductase**.

Hormonal contraceptives

Human fertility control currently rests firmly on the use by females of hormonal contraceptives. These agents can contain a mixture of synthetic oestrogens and progestogens (analogues of progesterone), or progestogens only, and are administered as daily tablets, depot injections that last for several months, or as long-term (5 years) uterine implants. They act as negative feedback signals to suppress gonadotrophin and thus prevent ovulation and/or conception. Male contraceptives based on the same principles are not yet available, although such drugs are under development.

47 Sexual differentiation and function

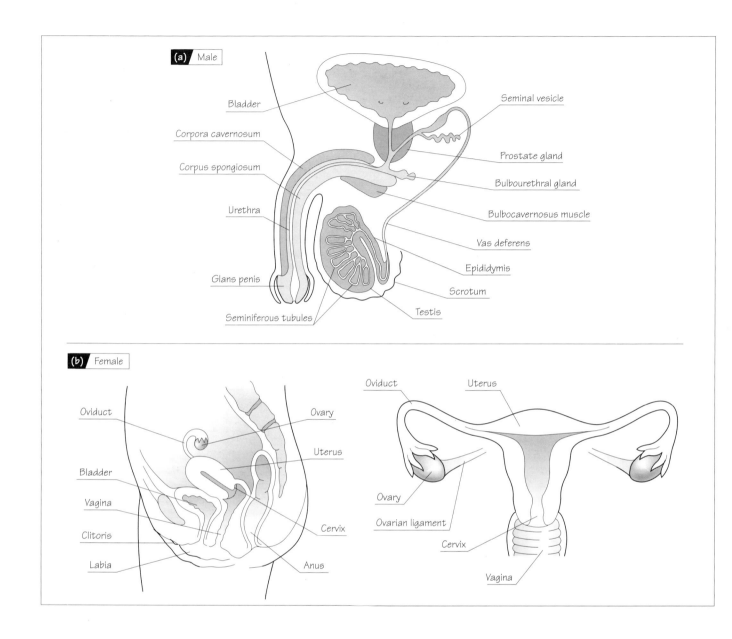

Table 47 Primary and secondary sexual characteristics that develop during puberty

Females (stimulated by oestrogens)	Males (stimulated by [dihydro]testosterone)
Growth and maturation of ovaries	Growth and maturation of testes, including descent into scrotum
Growth of external genitalia and pubic hair	Growth of external genitalia and pubic hair
Growth of breasts	Increased size of larynx, leading to deeper voice
Keratinization of vaginal mucosa, enlargement of uterus	Increased bone mass
Deposition of fat around hips and thighs	Increased muscle mass and strength
	Thickened skin
	Increased and thickened body hair

Sexual differentiation

Gender is determined by the presence of X and Y chromosomes in the genome (Chapter 1). Two X chromosomes provide the female genotype, whereas X and Y chromosomes together give a genetic male. The gonads are apparent after about 4 weeks of gestation. The *Sry* gene on the Y chromosome is thought to be responsible for the establishment of testicular development in males. The early gonads secrete steroids just as they do in the adult, and these hormones determine the sexual phenotype. Testosterone stimulates the development of the male genitalia (Fig. 47a) and the organization of neuronal systems in the brain that are involved in sexual function and behaviour. Notably, there is marked growth in the **sexually dimorphic nucleus** of the medial preoptic area of the hypothalamus and in the spinal nucleus that controls the **bulbo-cavernosus** muscle which is involved in ejaculation. Curiously, testosterone has to be converted to oestrogen by brain aromatases to have these effects. Although the male fetus is exposed to maternal oestrogens via the placenta, a binding protein in the fetal plasma effectively neutralizes any effects these may have. In the absence of the *Sry* gene (and consequent absence of testosterone), the development of reproductive organs and brain connectivity defaults to a female pattern which is dependent on the secretion of ovarian oestrogens. Thus, the steroid environment of the fetus determines the sexual phenotype of the adult.

Puberty

Although active before birth, the gonadotrophic axis quickly becomes quiescent after parturition and remains so until the onset of puberty at 12–14 years. The signal that triggers this event remains obscure, but it may result from an endogenous activation of brain pattern generating circuits that stimulate gonadotrophin-releasing hormone (GnRH) neurones. **Body mass**, signalled via circulating levels of leptin (Chapter 39) and insulin-like growth factor-1 (IGF-1) (Chapter 42), is an important permissive factor in females. The onset of menstruation (**menarche**) occurs at roughly a body mass of 47 kg irrespective of age, and undernutrition is associated with failure of the menstrual cycle. Puberty begins with increased release of luteinizing hormone (LH), first at night and then throughout the day. In males, this stimulates the release of testosterone from Leydig cells and the subsequent onset of spermatogenesis, accompanied by the many physical changes associated with the final growth into manhood (Table 47). It is believed that the appearance of secondary sexual characteristics is stimulated by the testosterone metabolite, dihydrotestosterone. In females, the onset of the cyclic release of LH gives rise to the beginning of menstruation and the development of the mature female body pattern (Table 47). The end of puberty marks the onset of full sexual maturity and the conclusion of somatic growth (Chapter 43). Figure 47a,b shows the mature male and female reproductive tracts.

Sexual function

Sexual attraction and behaviour in humans are the highly complex result of physiological factors, combined with societal and other psychological influences. The overall level of **libido** (sexual motivation) is set by the hypothalamus under the influence of higher centres and the hormonal environment. In males, sexual arousal arises from physical stimulation of the genitalia (a spinal reflex) or from psychological stimuli (by pathways descending from the hypothalamus via the brain stem) that activate sacral parasympathetic nerves (Chapter 8). The penis becomes erect as the result of the dilation of blood vessels entering the **corpora cavernosum** (the main erectile tissue) and **corpus spongiosum** (Fig. 47a). The enhanced flow of blood into the cavernous spaces increases tissue pressure and restricts venous drainage, causing a further build-up of pressure to make the penis fully erect. The parasympathetic nerves cause vasodilatation by the release of acetylcholine, vasoactive intestinal peptide and, primarily, **nitric oxide** (**NO**; Chapter 17). NO increases the manufacture of **cyclic guanosine monophosphate** (**cGMP**) in blood vessel smooth muscle cells to cause them to relax. Sildenafil (Viagra™) inhibits the breakdown of cGMP and thus enhances erectile function. The female sexual response sometimes involves erection of the clitoris, but the main manifestations are relaxation of the smooth muscles of the vagina and an increase in mucous secretions that act as a lubricant. Again, these actions are brought about by the activation of parasympathetic nerves. The combined effects of the male and female sexual responses facilitate entry of the penis into the vagina (**intromission**). Frictional forces stimulate mechanoreceptors in the glans penis and the clitoris that eventually lead to reflex activation of the sympathetic nerves that causes **orgasm**. In the male, this involves peristaltic contractions of the epididymis to pump sperm into the urethra, where they are mixed with the secretions of the **bulbourethral gland**, the **seminal vesicle** and the **prostate gland** to form semen. The secretions provide, respectively, lubrication, energy (in the form of the sugar **fructose**) and an alkaline barrier against the acid conditions normally prevalent in the vagina. They also include high levels of the arachidonic acid-derived local hormones **prostaglandins** (Chapter 9) that stimulate the motility of sperm and of the female tract. Further peristaltic contractions of the urethra, in combination with the action of the bulbocavernosus muscle, emit the semen bolus into the upper end of the vagina (**ejaculation**). The female orgasm, which may involve the release of pituitary **oxytocin** elicited by mechanical stimulation of the cervix (Chapter 48), results in rhythmic contractions of the vaginal and uterine muscles that promote the flow of semen into the uterus. Sperm move by means of their own motility and by the beating of cilia on the walls of the uterus, but only a few hundred sperm of the millions released in a single ejaculate will complete the 6-hour journey from the vagina to the oviducts.

48 Fertilization, pregnancy and parturition

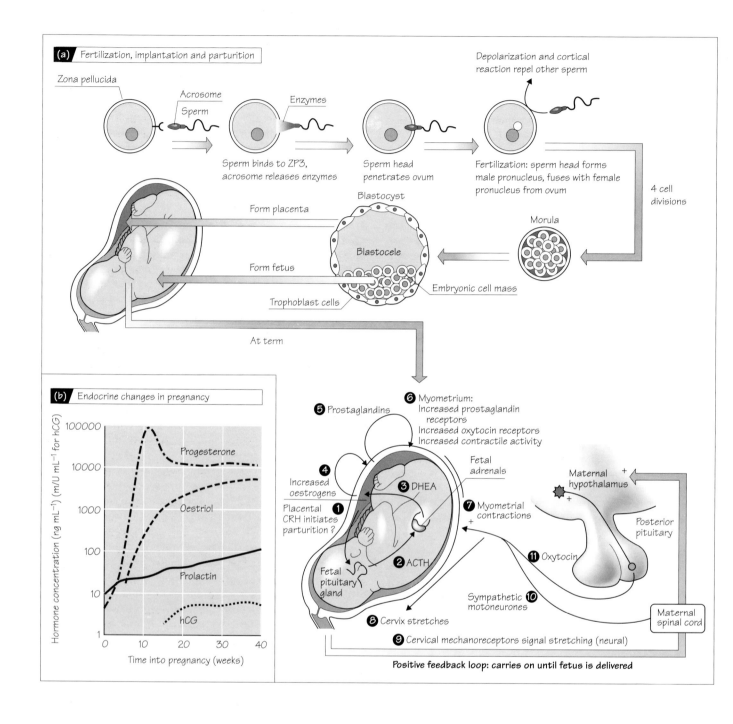

(a) Fertilization, implantation and parturition

Zona pellucida

Acrosome
Sperm

Enzymes

Depolarization and cortical
reaction repel other sperm

Sperm binds to ZP3,
acrosome releases enzymes

Sperm head
penetrates ovum

Fertilization: sperm head forms
male pronucleus, fuses with female
pronucleus from ovum

4 cell
divisions

Form placenta

Blastocyst

Morula

Blastocele

Form fetus

Embryonic cell mass

Trophoblast cells

At term

(b) Endocrine changes in pregnancy

Hormone concentration (ng mL⁻¹) (m/U mL⁻¹ for hCG)

Progesterone

Oestriol

Prolactin

hCG

Time into pregnancy (weeks)

❺ Prostaglandins

❻ Myometrium:
Increased prostaglandin
receptors
Increased oxytocin receptors
Increased contractile activity

❹
Increased
oestrogens

❸ DHEA

Fetal
adrenals

Maternal
hypothalamus

❶
Placental
CRH initiates
parturition?

❼ Myometrial
contractions

Posterior
pituitary

❷ ACTH

⓫ Oxytocin

Fetal
pituitary
gland

Sympathetic
motoneurones ⓾

Maternal
spinal cord

❽ Cervix stretches

❾ Cervical mechanoreceptors signal stretching (neural)

Positive feedback loop: carries on until fetus is delivered

Fertilization

The unfertilized ovum can survive for 1–2 days after ovulation, and sperm remain viable in the uterus for about the same time after ejaculation. HCO_3^- ions in the uterus trigger the **capacitation** of sperm. This is a prerequisite for fertilization that involves remodelling of the lipids and glycoproteins of the sperm plasma membrane, coupled with increased metabolism and motility. The ovum is surrounded by the **zona pellucida**, an acellular membrane bearing the glycoprotein **ZP3** that acts as a sperm receptor. Fertilization occurs in the oviduct, when a single capacitated sperm binds to ZP3 and undergoes the **acrosome reaction**. The acrosome is a body containing proteolytic enzymes that is attached to the sperm head (Fig. 48a). When a sperm binds to ZP3, the acrosomal enzymes are released to digest a pathway for the sperm to penetrate the ovum, within which the contents of the sperm head, including its genetic material, are deposited. This event leads to a chain of reactions that denies access to further sperm penetration. The ovum first undergoes electrical depolarization and then discharges granules that impair further sperm binding at the zona pellucida (the **cortical reaction**). In this way, fertilization is normally restricted to one sperm per ovum. Some 2–3 h after penetrating the ovum, the sperm head forms the **male pronucleus** which joins with the **female pronucleus** from the ovum (Fig. 48a). Fusion of the pronuclei combines the parental genetic material from the gametes to form the **zygote**.

Pregnancy

The zygote is propelled by ciliary action into the uterus, where it must implant itself in the endometrium. During this journey, the zygote undergoes a number of cell divisions to form the **morula**, a solid ball of 16 cells that 'hatches' from the zona pellucida and develops into the **blastocyst**, in which embryonic cells are surrounded by **trophoblasts** (Fig. 48a). The trophoblasts are responsible for implantation, digesting away the uterine endometrial wall to form a space for the embryo, opening up a pathway to the maternal circulation (via the **spiral arteries** of the uterus) and forming the fetal portion of the **placenta**. The tissue engineering activities of trophoblasts are mediated by epidermal growth factor (EGF) (Chapter 42) and interleukin-1β (Chapter 10). Implantation is complete within 7–10 days of fertilization, at which time the embryo and early placenta begin to secrete **human chorionic gonadotrophin (hCG)**. The appearance of hCG in the plasma and urine is one of the earliest signs of successful conception, and its detection forms the basis of pregnancy testing kits. hCG is a glycoprotein similar to LH that stimulates progesterone secretion, initially from the corpus luteum (which is maintained for a while by hCG) and later by the placenta. Progesterone levels rise steadily throughout pregnancy and fall sharply at term (Fig. 48b). This steroid ensures that the smooth muscle of the uterus remains quiescent during gestation (essential for a successful pregnancy), stimulates mammary gland development and prepares the maternal brain for motherhood. The placenta also secretes **chorionic somatomammatrophin**, a growth hormone-like peptide that mobilizes metabolic fuels (Chapter 39) and promotes mammary gland growth, and oestrogen (mainly **oestriol**) that stimulates uterine expansion to accommodate the growing embryo. Fetal development occurs within a fluid-filled sac, known as the **amniotic membrane**, which provides a protective buffer against physical trauma. Pregnancy makes many physiological demands on the mother. The ventilation rate, cardiac output and plasma volume increase to supply fetal–maternal oxygen and water demands; the gastrointestinal absorption of minerals is enhanced; and the renal glomerular filtration rate (Chapter 28) rises to cope with fetal waste production.

Parturition

After some 40 weeks of gestation, the fetus is ready for life outside the uterus. The signal that initiates parturition is still not fully understood, and appears to be different between primates and lower mammals. In primates, the primary signal is thought to arise from the **fetoplacental unit** (i.e. the fetus plus the placenta) as an increase in dehydroepiandrosterone (DHEA) production from the fetal adrenal cortex (Chapter 45), which may be driven by *placental* (rather than hypothalamic) production of corticotrophin-releasing hormone (CRH) (Chapters 40 & 45; Fig. 48a). DHEA is a precursor for oestrogen production by the placenta. As the placental aromatase enzymes are not rate limiting, an increase in DHEA, which is a precursor of oestrogen (Chapter 46), automatically increases oestrogen production. Whatever the initiating signal might be, the end result is an increase in the synthesis of prostaglandins E and F (Chapter 9) by fetal and uterine tissues, with concomitant increases in prostaglandin receptors in the uterine smooth muscle. The prostaglandins stimulate the production of uterine receptors for **oxytocin** and change the pattern of activity in the uterine myometrium from slow, gentle contractions to regular, deep contractions that eventually move the fetus into the cervix. The cervix, which is softened by prior release of the ovarian peptide hormone **relaxin**, dilates as the fetus is forced downwards. At this time, the amniotic membrane ruptures. Stretching of the cervix activates mechanoreceptors that stimulate a spinal sympathetic reflex which causes myometrial contraction and secretion of **oxytocin** from the posterior pituitary gland (Chapter 40; Fig. 48a). Oxytocin is a powerful stimulant of uterine smooth muscle that causes further contraction of the myometrium and pushes the fetus further into the cervix, resulting in further stimulation of mechanoreceptors and leading to the release of more oxytocin, i.e. this is a positive feedback system. The spinal reflex, aided by waves of oxytocin, generates large, regular contractions of the uterus that eventually expel the fetus and placenta through the vagina, completing the birth process. Oxytocin continues to be useful, as it limits maternal bleeding by causing vasoconstriction. In the fetus, oxytocin closes the **ductus arteriosus**, a blood vessel that shunts blood away from the pulmonary circulation *in utero*, but which would obviously hamper postpartum life should it remain open.

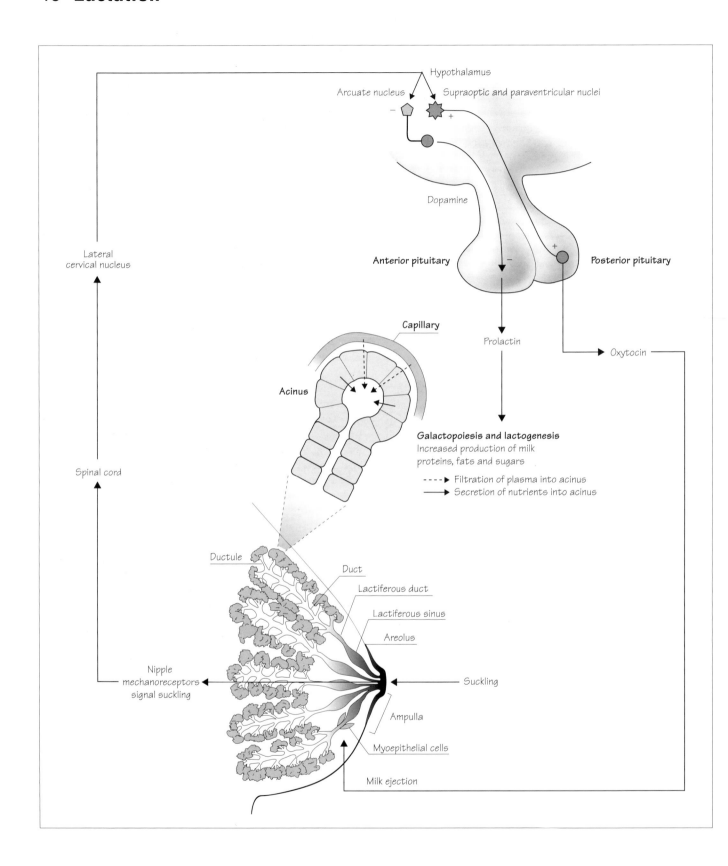

Milk, which sustains mammalian infants through the first few months of life, is produced by the mammary glands (Fig. 49) under the influence of the pituitary protein hormone **prolactin** (Chapter 40). The glands comprise several lobules that are composed of **acini**, similar in structure to the salivary glands and the exocrine pancreas (Chapters 33 & 36). The lobules empty into **lactiferous ducts**. As the ducts approach the **areola** (nipple), they open out to form **lactiferous sinuses** before narrowing again to emerge at the **ampulla** on the nipple. The ducts and sinuses are organized to collect milk rather than allowing a free flow to the ampulla. They are lined by **myoepithelial cells** that contract to expel milk from the breast. Progesterone, oestrogen, prolactin, cortisol and growth hormone are all required for the complete development of the mammary glands, which is achieved only at the time of giving birth; for the rest of adult life the glandular tissue is rather small. Milk is formed by the filtration of plasma into the acinus. The acinar and duct cells then modify the plasma by the addition of fats (triglycerides), proteins (principally **casein**, α-lactalbumin and lactoglobulin B) and sugars (mostly **lactose**) to produce an isotonic liquid that is roughly 4% fat, 1% protein and 7% sugar, with almost 100 additional trace nutrients, including many ions (including Ca^{2+}), some immunoglobulins in the form of IgA (Chapter 10) and growth factors, such as insulin-like growth factor-1 (IGF-1) and epidermal growth factor (EGF) (Chapter 42). **Colostrum**, the first secretion of the mammary glands after birth, is particularly rich in protein, but has a lower sugar concentration than mature milk. It also contains high levels of **antibodies** (Chapter 10) that provide the infant with basic immunological protection in the first days of life.

Plasma prolactin levels rise steadily during pregnancy, but the lactogenic effects of the hormone are inhibited by the presence of progesterone and oestrogen, so that its main role during gestation is to promote mammary growth. The loss of placental steroids at term (Chapter 48) allows prolactin to exert its full effects on milk production, provided that cortisol and insulin are also present. However it should be noted that without pre-exposure to progesterone and oestrogen, the mammary glands do not respond to prolactin. The hormone acts through a receptor linked to a Janus kinase–signal transduction and activation of transcription (JAK–STAT) system (Chapter 43) that activates the genes producing milk proteins and the synthetic enzymes for lactose and triglycerides. The production of nutrients is termed **galactopoiesis**. Prolactin also increases blood flow to the gland, and stimulates the delivery of nutrients into milk by exocytosis (proteins) or specific membrane transport systems (sugars, fats, antibodies); these actions are referred to as **lactogenesis**.

Prolactin is an unusual anterior pituitary hormone in that it is released **constitutively** (i.e. without a stimulus) from pituitary lactotrophs, and the primary control from the hypothalamus is inhibitory via **dopamine**, although other hypophysiotropic hormones may also be involved (Chapter 40; Fig. 49). After birth, the main stimulus that maintains prolactin release is **suckling**. Milk production thus continues for as long as the infant continues to feed from the mother. Prolactin inhibits luteinizing hormone (LH) release from the pituitary and maintains the mother in a low state of fertility until the infant is weaned. This is a useful mechanism for spacing out births, but is not 100% effective in humans. Prolactin-secreting tumours of the pituitary render the patient infertile, but this can be overcome by the administration of the dopamine agonist bromocriptine, which inhibits prolactin release long enough for ovulation to occur. Prolactin is released in several conditions other than around birth: sleep, stress, eating and exercise are all associated with elevated plasma prolactin, although the exact function of this release is not yet known.

Milk let down reflex

Prolactin stimulates milk production, but another hormone is required to eject milk onto the surface of the nipple. Stimulation of areolar mechanoreceptors by suckling infants activates a neural pathway that ascends to the **paraventricular** and **supraoptic nuclei** of the hypothalamus via the lateral cervical nucleus of the brain stem. This pathway excites magnocellular neurones (Chapter 40) to secrete pulses of oxytocin into the blood at 10–20 min intervals. It is not certain how the suckling stimulus, which is continuous, is translated into episodic activity in oxytocin-releasing cells. The oxytocin pulses seem to arise from the simultaneous activation of all oxytocin neurones in both nuclei. The hormone is a potent stimulant of myoepithelial cells, which pump milk from the lactiferous sinuses out through the nipple and into the mouth of the infant. Milk let down encourages further suckling by the recipient, which leads to more oxytocin release, and so makes up another positive feedback system that operates until the infant is sated. This **milk ejection reflex** (Fig. 49) is also stimulated in response to the crying of infants as a result of psychological **conditioning**. However, the reflex is strongly inhibited by maternal stress, which is one of the most common causes of failure of lactation in new mothers. In animals, the release of oxytocin in the brain has been shown to facilitate maternal behaviour, but works only after pre-exposure to progesterone and oestrogens.

50 Skeletal muscle and its contraction

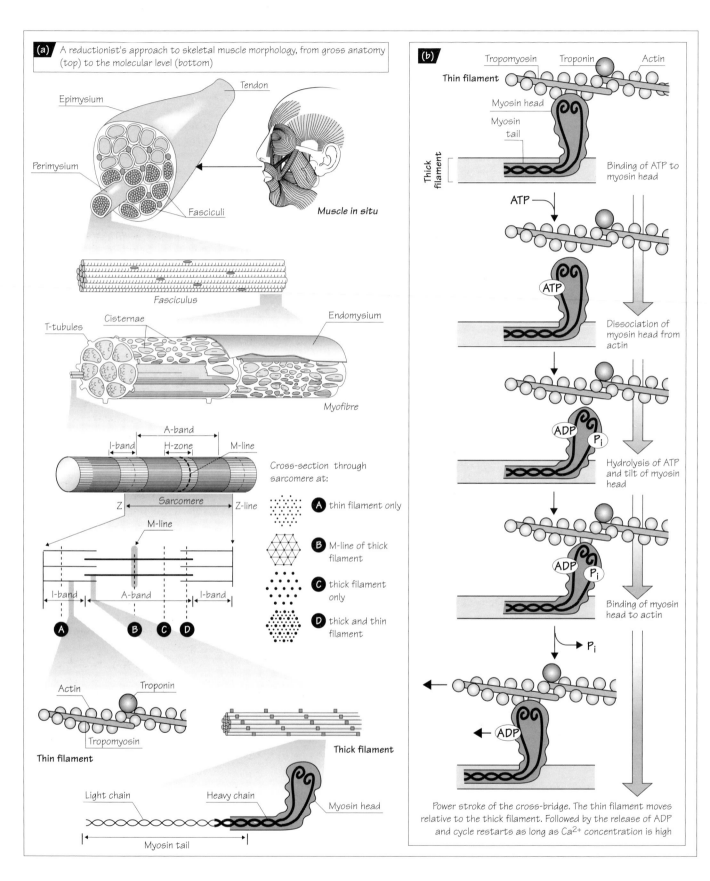

Muscles make up about 50% of the adult body mass. There are three types of muscle: **skeletal** (muscle attached to the skeleton); **cardiac** (muscle involved in cardiac function) (see Chapter 53) (both of these are morphologically striated or striped and are commonly called **striated muscles**); and **smooth** (muscle involved in many involuntary processes in the blood vessels and gut; this type is not structurally striated, hence its name) (see Chapter 53). A comparison of the properties of the three muscle types is shown in the Appendix (p. 130).

Skeletal muscle

The **skeletal muscles** and the **skeleton** function together as the musculoskeletal system. Skeletal muscle is sometimes referred to as **voluntary muscle** because it is under conscious control. It uses about 25% of our oxygen consumption at rest and this can increase up to 20-fold during exercise.

General mechanisms of skeletal muscle contraction

The functions of muscle tissue are the development of tension and shortening of the muscle. Muscle fibres have the ability to shorten a considerable amount, which is brought about by the molecules sliding over each other. Muscle activity is transferred to the skeleton by the tendons, and the tension developed by the muscles is graded and adjusted to the load.

Fine structure of skeletal muscle (Fig. 50a)

The connective tissue surrounding the whole muscle is called the **epimysium**. The connective tissue that extends beyond the body of the muscle eventually blends into a **tendon**, which is attached to bone or cartilage. Skeletal muscle is composed of numerous parallel, elongated, multinucleated (up to 100) cells, referred to as **muscle fibres** or **myofibres**, which are between 10 and 100 μm in diameter and vary in length, and are grouped together to form **fasciculi**. Each fasciculus is surrounded by the **perimysium**. Each myofibre, which is surrounded by the **endomysium**, is made up of 1 μm diameter **myofibrils** separated by cytoplasm and arranged in a parallel fashion along the long axis of the cell. Each myofibril is further subdivided into **thick** and **thin myofilaments** (thick, 10–14 nm in width and 1.6 μm in length; thin, 7 nm in width and 1 μm in length). These are responsible for the cross-striations. **Thin filaments** consist primarily of three proteins, **actin**, **tropomyosin** and **troponin**, in the ratio 7 : 1 : 1, and **thick filaments** consist primarily of **myosin**. The cytoplasm surrounding the **myofilaments** is called the **sarcoplasm**. Each myofibre is divided at regular intervals along its length into **sarcomeres** separated by **Z-discs** (in longitudinal sections, these are **Z-lines**). To the Z-lines are attached the thin filaments held in a hexagonal array. The **I-band** extends from either side of the Z-line to the beginning of the thick filament (myosin). The myosin filaments make up the **A-band**.

The **H-zone** is at the centre of the sarcomere, and the **M-line** is a disc of delicate filaments in the middle of the H-zone that holds the myosin filaments in position so that each one is surrounded by six actin filaments.

The thin filaments consist of two intertwining strands of actin with smaller strands of tropomyosin and troponin between the intertwining strands. Each strand of actin is made up of about 200 units of globular or G-actin. It is on these globules that there is a site for myosin to bind during contraction.

The thick filaments are made up of about 100 myosin molecules; each molecule is club shaped, with a thin tail (shaft) comprising two coiled light peptide chains and a head made up of two heavy peptide chains and four light peptide chains that have a regulatory function. The ATPase activity of the myosin molecule is concentrated in the head.

The thin tails of the myosin molecules form the bulk of the thick filaments, whereas the heads are 'hinged' and project outward to form cross-bridges between the thick filaments and their neighbouring thin filaments. Six thin filaments surround each thick filament.

Between the myofibrils are a large number of mitochondria and glycogen granules, as found in other cells, but muscle cells have regular invaginations which project from outside the cell and wrap around the sarcomeres, particularly where the thin and thick filaments overlap. These invaginations are called transverse or **T-tubules** and contain extracellular fluid. The specialized smooth endoplasmic reticulum, the **sarcoplasmic reticulum**, which is close to the T-tubules, is enlarged to form **terminal cisternae** which actively transport Ca^{2+} into the lumen from the sarcoplasm.

Like fingers of the hands sliding over one another, actin and myosin molecules slide past each other. The myosin heads bind to the actin chain and tilt. There is a constant process of binding, tilting, releasing and rebinding of cross-bridges, as well as rotation of the myosin filaments as they interact with the actin filaments and bind with the alternate myofibril in the hexagonal structure. This results in the contraction of the whole muscle. The cross-bridges are formed asynchronously so that some are active, whilst others are resting.

The interaction of actin (thin filaments) and myosin (thick filaments) brings about contraction of the muscle, which is caused by the cross-bridges, a result of the interaction of troponin and Ca^{2+}. This mechanism is called the **sliding filament theory**. The contraction of muscle is triggered by the release of Ca^{2+} from the sarcoplasmic reticulum. Ca^{2+} floods out of the cisternae, where it is stored by binding reversibly with a protein, **calsequestrin**. This raises the concentration of calcium from $0.1\,\mu mol\,L^{-1}$ to more than $10\,\mu mol\,L^{-1}$, saturating the binding sites on troponin. This results in a shift of tropomyosin, thus allowing the myosin cross-bridges to bind more strongly with actin and begin the contraction cycle (Fig. 50b). The heads tilt after attachment by hydrolysing the adenosine triphosphate (ATP) energy stores, releasing adenosine diphosphate (ADP) and inorganic phosphate (P_i), which leads to a greater binding of the cross-bridges. ADP and P_i escape from the head, freeing the head for another molecule of ATP. This releases the binding of the head and, if Ca^{2+} is still present, the cycle continues. Otherwise, the binding is inhibited. Contraction is maintained as long as Ca^{2+} is high. The duration of the contraction is dependent on the rate at which the sarcoplasmic reticulum pumps back the Ca^{2+} into the terminal cisternae.

51 Neuromuscular junction and whole muscle contraction

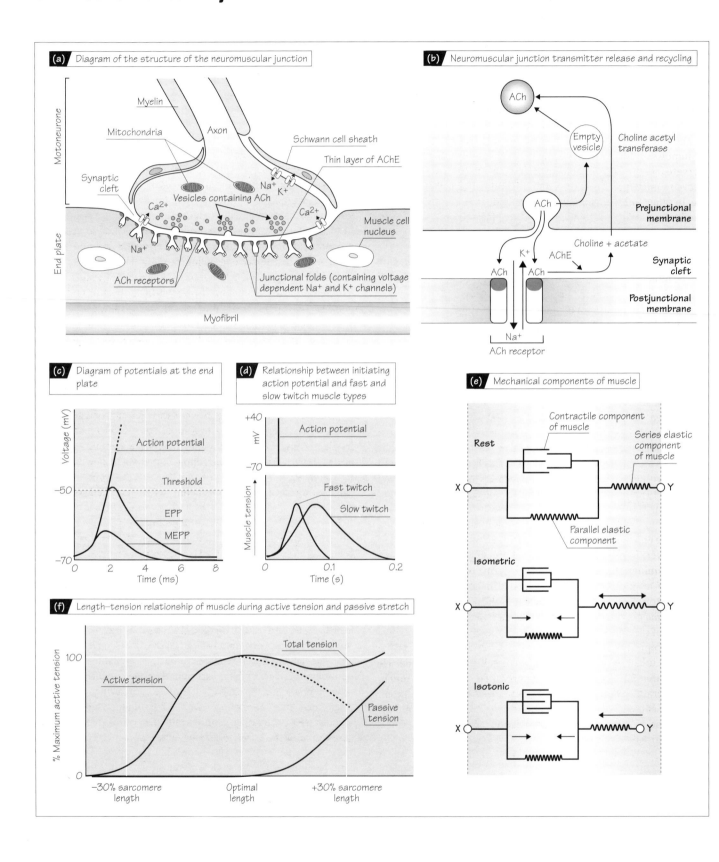

(a) Diagram of the structure of the neuromuscular junction

(b) Neuromuscular junction transmitter release and recycling

(c) Diagram of potentials at the end plate

(d) Relationship between initiating action potential and fast and slow twitch muscle types

(e) Mechanical components of muscle

(f) Length–tension relationship of muscle during active tension and passive stretch

Neuromuscular junction

For **skeletal (voluntary) muscle** to contract, there must be activation from higher centres in the brain to the muscle fibres themselves. The neurones that innervate skeletal muscles are called **α-motor neurones**. Each motor nerve splits into a number of branches that make contact with the surface of individual muscle fibres in the form of bulb-shaped endings. These endings are arranged in groups and make connections with a specialized structure on the surface of the muscle fibre, called the **motor end plate**, and together form the **neuromuscular junction (NMJ)** (Fig. 51a).

The role of the NMJ is the one-to-one transmission of excitatory impulses from the α-motor neurone to the muscle fibres it innervates. It forms a reliable transmission of the impulses from nerve to muscle and produces a predictable response in the muscle. In other words, an action potential in the motor neurone must produce an action potential in the muscle fibres it innervates; this, in turn, must produce a contraction of the muscle fibres. The process by which the **NMJ** produces this one-to-one response is shown in Figs 51a and b.

The motor neurone axon terminal has a large number of vesicles containing the transmitter substance **acetylcholine (ACh)**. At rest, when not stimulated, a small number of these vesicles release their contents, by a process called **exocytosis**, into the synaptic cleft between the neurones and the muscle fibres. ACh diffuses across the cleft and reacts with specific ACh receptor proteins in the postsynaptic membrane (motor end plate). These receptors contain an integral ion channel, which opens and allows the movement inwards (influx) of small cations, mainly Na^+. There are more than 10^7 receptors on each end plate (**postjunctional membrane**); each of these can open for about 1 ms and allow small positively charged ions to enter the cell. This movement of positively charged ions generates an **end plate potential (EPP)**. This is a depolarization of the cell with a rise-time of approximately 1–2 ms and may vary in amplitude (unlike the all-or-nothing response seen in the action potential). The random release of ACh from the vesicles at rest gives rise to small, 0–4 mV depolarizations of the end plate, called **miniature end plate potentials (MEPP)** (Fig. 51c).

However, when an **action potential** reaches the **prejunctional nerve terminal**, there is an enhanced permeability of the membrane to Ca^{2+} ions. This causes an increase in the exocytotic release of ACh from several hundred vesicles at the same time. This sudden volume of ACh diffuses across the cleft and stimulates a large number of receptors on the postsynaptic membrane, and thus produces an epp that is above the threshold for triggering an action potential in the muscle fibre. It triggers a **self-propagating muscle action potential**. The depolarizing current (the **generator potential**) generated by the numbers of quanta of ACh is more than sufficient to cause the initiation of an action potential in the muscle membrane surrounding the postsynaptic junction. The typical summated **EPP** is usually four times the potential necessary to trigger an action potential in the muscle fibre, and so there is a large inherent safety factor.

The effect of ACh is rapidly abolished by the activity of the enzyme **acetylcholinesterase**. ACh is hydrolysed to **choline** and **acetic acid**. About one-half of the choline is recaptured by the presynaptic nerve terminal and used to make more ACh. Some ACh diffuses out of the cleft, but the enzyme destroys most of it. The number of vesicles available in the nerve ending is said to be sufficient for only about 2000 nerve–muscle impulses, and therefore the vesicles reform very rapidly within about 30 s (Fig. 51b).

Whole muscle contraction

As the action potential spreads over the muscle fibre, it invades the T-tubules and releases Ca^{2+} from the **sarcoplasmic reticulum** into the **sarcoplasm**, and the muscle fibres that are excited contract. This contraction will be maintained as long as the levels of Ca^{2+} are high. The single contraction of a muscle due to a single action potential is called a **muscle twitch**. Fibres are divided into **fast** and **slow twitch fibres** depending on how fast the sarcoplasmic reticulum can pump Ca^{2+} back into the **terminal cisternae**. Fast twitch fibres can achieve this in about 10 ms, whilst slow twitch fibres can take up to 80 ms. Different muscles are made up of different proportions of these two types of fibre, leading to a huge variation in overall muscle contraction times. Figure 51d shows a diagram of fast and slow twitch muscle types.

At rest, muscles exert tension when stretched. Muscles have a passive elastic property and act both in series and parallel to the contractile element (Fig. 51e).

Isometric contraction occurs when the two ends of a muscle are held at a fixed distance apart, and stimulation of the muscle causes the development of tension within the muscle without a change in muscle length. **Isotonic contraction** occurs when one end of the muscle is free to move and the muscle shortens whilst exerting a constant force. In practice, most contractions are made up of both elements.

The relationships between resting, active and total tensions developed in skeletal muscle are shown in Fig. 51f. The **passive curve** is due to the stretching of the elastic components, the **active curve** is due to contraction of the sarcomeres alone (contractile component), and the **total curve** is due to the sum of the passive and active tensions developed. It can be seen that the active tension developed is dependent on the length of the muscle. The optimum length occurs where the thick and thin filaments are thought to provide a maximum number of active cross-bridge sites for interaction (this length is very close to that of the resting length of a particular muscle). As the muscle length is increased, the thick and thin filaments overlap less, providing less cross-bridge sites for interaction; as the muscle shortens, the thin filaments overlap one another and, in so doing, reduce the number of active sites available for interaction with the thick filaments.

52 Motor units, recruitment and summation

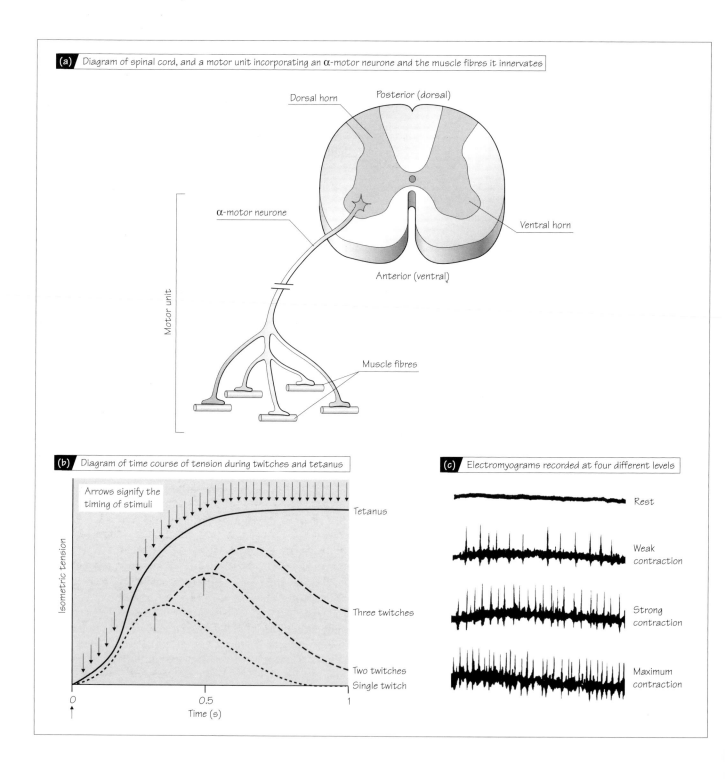

(a) Diagram of spinal cord, and a motor unit incorporating an α-motor neurone and the muscle fibres it innervates

Dorsal horn

Posterior (dorsal)

α-motor neurone

Ventral horn

Anterior (ventral)

Motor unit

Muscle fibres

(b) Diagram of time course of tension during twitches and tetanus

Arrows signify the timing of stimuli

Isometric tension

Tetanus

Three twitches

Two twitches

Single twitch

0   0.5   1

Time (s)

(c) Electromyograms recorded at four different levels

Rest

Weak contraction

Strong contraction

Maximum contraction

In normal skeletal muscle, fibres never contract as isolated individuals. Several contract at almost the same time, as they are all supplied by the same α-motor neurone. The single motor neurone and all the fibres it innervates is called the **motor unit** (Fig. 52a). This is the smallest part of a muscle that can be made to contract independently of other parts of the muscle. The number of muscle fibres innervated by one motor unit can be as low as two to three or as high as over 2000. The number is correlated with the precision with which the tension developed by the muscle is graded.

Within the muscle, the muscle fibres of each motor unit are widely distributed amongst the fibres of many other units. This, in effect, distributes the demands made on the muscle's circulatory support. The ratio between the number of α-motor neurones and the total number of skeletal muscle fibres is small in muscles such as the extraocular muscles that involve fine smooth movements (1:5), but large in muscles such as the gluteus maximus that need to generate powerful but coarse movements (1:>1000).

Motor unit fibres can be classified into three types: **I, IIA** and **IIB. Type I** are small-diameter, slow units, with fatigue-resistant, relatively weak fibres; they have a rich capillary supply, and contain mostly oxidative metabolic processes. **Type IIA** are medium-to small-diameter, fast units, with fatigue-resistant, intermediate strength fibres; they have a rich capillary supply, and contain both oxidative and glycolytic metabolic processes. **Type IIB** are large-diameter, fast units that are easily fatigued; they are however strong fibres, with a relatively sparse capillary supply, and contain mostly glycolytic metabolic processes.

Most muscles contain all three types, but differ in the proportions of each according to the function of the muscle as a whole. Posture muscles, such as the soleus, have mostly slow Type I units, whereas movement muscles, such as the gastrocnemius, have a high proportion of Type IIA and IIB units. Training and exercising can alter these proportions.

The cell bodies of α-motor neurones also vary in size according to the type of motor unit: **motor neurones innervating Type I fibres have the smallest cell bodies, and those innervating Type IIB fibres have the largest.**

During graded contraction, there is a recruitment order of the units, such that the smallest cells discharge first and the largest last (the so-called **size principle**). **Force** is controlled not only by **varying the unit recruitment**, but also by **varying the firing rate of the units**. A single action potential in a single motor unit produces a delayed rise in tension in all the muscle fibres that make up that motor unit. A second and a third action potential that occur soon after the first produce a summated contraction, or series of twitches. The tension developed by the first action potential has not completely decayed when the second contraction is grafted onto the first, and so on for the third action potential and contraction. This is called **summation** (Fig. 52b). If the muscle fibres are stimulated repeatedly at a faster frequency, a sustained contraction results in which individual twitches cannot be detected. This is called **tetanus**. The tension of tetanus is much greater than the maximum tension of a single, double or triple twitch (Fig. 52b). For most units, the firing rate for a steady contraction is between 5 and 8 Hz. It can rise to 40 Hz or more, but only for very brief periods. During a gradual increase in contraction of a muscle, the first units start to discharge and increase their firing rate and, as the force needs to increase, new units are recruited and, in turn, also increase their firing rate. When there is a need to gradually decrease the force output, the pattern is reversed, so that those units that were recruited last will be the first to decrease their firing and then stop, and the last units to fire will be the smallest units. Because the unitary firing rates for each motor unit are different and not synchronized, the overall effect is a smooth force profile from the muscle. The greater the desynchronized firing, the smoother the movements observed. When synchronized firing does occur, such as in fatigued states and Parkinson's disease, marked muscle tremors are seen.

The summated excitatory impulses (action potentials) of the motor units can be recorded in an **electromyogram** (**EMG**). The EMG is an extracellular recording made from either the skin surface overlying a muscle or from electrodes inserted extracellularly within the body of the muscle. The increase in recruitment of individual motor units, as well as the increased rate of firing of the units, can sometimes be seen in the EMG during increased force of contraction (Fig. 52c).

53 Cardiac and smooth muscle

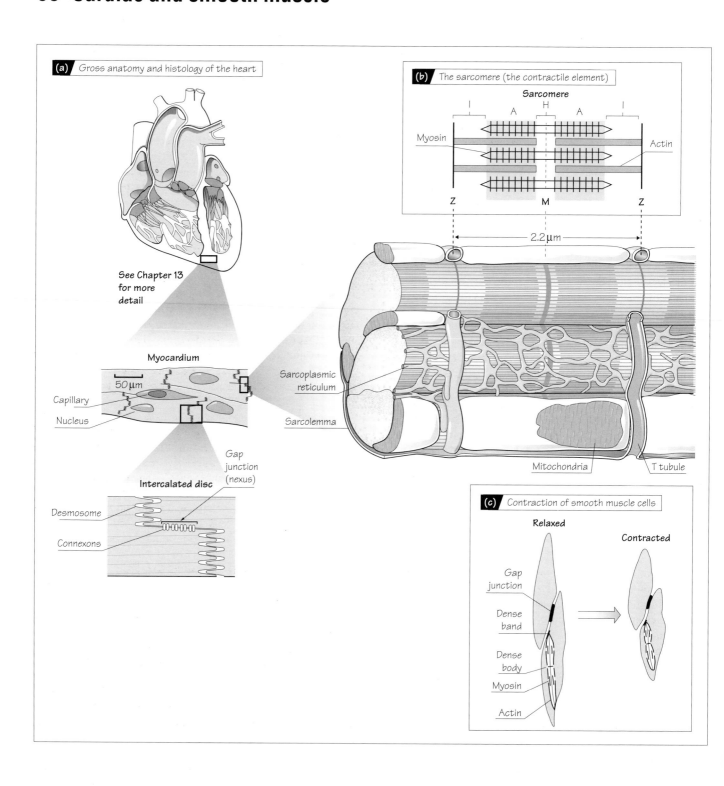

(a) Gross anatomy and histology of the heart

See Chapter 13 for more detail

Myocardium

50 μm

Capillary

Nucleus

Gap junction (nexus)

Intercalated disc

Desmosome

Connexons

(b) The sarcomere (the contractile element)

Sarcomere

I   A   H   A   I

Myosin

Actin

Z   M   Z

2.2 μm

Sarcoplasmic reticulum

Sarcolemma

Mitochondria

T tubule

(c) Contraction of smooth muscle cells

Relaxed

Contracted

Gap junction

Dense band

Dense body

Myosin

Actin

Cardiac muscle

The muscles of the heart, the **myocardium**, generate the force of contraction of the atrial and ventricular muscles. The myocardium is composed of cardiac muscle cells called **myocytes**. These cells are striated due to the orderly arrangement of the thick and thin filaments which, as in skeletal muscle, make up the bulk of the muscle. However, they are less organized than in skeletal muscle (Fig. 53a, b). The **myocytes** have dimensions of $100\,\mu m \times 20\,\mu m$, are branched, with a **single nucleus**, and are also **rich in mitochondria**. The normal pumping action of the heart is dependent on the synchronized contraction of all cardiac cells. Their contraction is not dependent on an external nerve supply, as in skeletal muscle, but instead the heart generates its own rhythm, called **inherent rhythmicity**. The nerves innervating the heart only speed up or slow down the rhythm and can modify the force of contraction (the so-called **chronotropic** and **inotropic** effects respectively; see Chapter 15).

The synchronicity between myocytes occurs because all the adjacent cells are linked to one another at their ends by **specialized gap** or **electrotonic junctions**—intercalated discs—which are essentially low-resistance pathways between cells. These allow action potentials to spread rapidly from one cell to another and enable the cardiac muscle to act as a **functional syncytium** (i.e. it acts as a single unit although comprising individual cells).

The intercalated discs provide both a structural attachment (**desmosomes**) between cells and an electrical contact, called a gap junction, made up of proteins called **connexons** (Fig. 53). Although a rise in intracellular $[Ca^{2+}]$ initiates contraction in the same way as in skeletal muscle (Chapter 50), the mechanisms leading to this rise in intracellular $[Ca^{2+}]$ are fundamentally different, and are discussed in Chapter 15.

Smooth muscle

The **absence of striations** within the cells and the poorer organization of the fibres give this type of muscle its name. Each cell contains only **one nucleus** situated near the centre. Smooth muscle is involved in many involuntary processes in blood vessels and the gut.

The smooth muscle of each organ is distinctive from that of most other organs, and there is considerable variation in the structure and function of smooth muscle in different parts of the body; however, essentially it can be divided into **unitary** (or **visceral smooth muscle**) and **multiunit** smooth muscle types.

Smooth muscle cells are spindle-shaped with dimensions of 50–$400\,\mu m$ in length by 2–$10\,\mu m$ thick. They are joined, like cardiac muscle, by special intercellular connections called **desmosomes**. Because the actin and myosin filaments are not regularly arranged, they lack striations. Smooth muscle cells shorten by sliding of the myofilaments towards and over one another, but at a much slower rate than in other muscle types. For this reason, they are capable of prolonged, maintained contraction, without fatigue and with little energy consumption (Fig. 53c).

The **unitary muscle type** or **visceral smooth muscle** exhibits many gap junctions between cells, and a steady wave of contraction can pass through a whole sheet of muscle as if it were a single unit. It is commonly found in the stomach, intestines, urinary bladder, urethra and blood vessels, and is capable of bringing about **autorhythmical activity** (seen particularly in the digestive tract where it is modulated by neuronal activity).

Tonic activity causes smooth muscle to remain in a constant state of contraction or tonus. It is commonly found in sphincters that control the movement of digestive products through the gastrointestinal tract.

Multiunit smooth muscle is made up of individual fibres not connected by gap junctions, but separately stimulated by autonomic motor neurones. Each smooth muscle fibre can contract independently from the others. Examples include the ciliary muscles of the eye, the iris of the eye and piloerector muscles that cause erection of hairs when stimulated by the sympathetic system.

The factors that influence the neural control of smooth muscle are:
1 The type of innervation and the transmitter released.
2 The receptor of the neurotransmitter on the muscle cell itself.
3 The anatomical arrangement of the nerve in relation to the muscle fibres.

There are three types of innervation: **extrinsic**—from the autonomic part of the nervous system, mainly sympathetic (arteries), parasympathetic (ciliary muscles) and both sympathetic and parasympathetic (gut); **intrinsic**—a plexus of nerves within the smooth muscle itself (seen in the gut); and **afferent sensory neurones**—these indirectly lead to the reflex activation of motor neurones.

Smooth muscle cells also respond to local tissue factors and hormones, i.e. changes in the fluids that surround them (interstitial fluids). In addition, many hormones that circulate in the bloodstream also cause smooth muscle contraction (hormones such as adrenaline (epinephrine), angiotensin, oxytocin, antidiuretic hormone (ADH), noradrenaline (norepinephrine) and serotonin). Also, a lack of oxygen in the tissues causes smooth muscle cells to relax and vasodilate; an increase in CO_2 or H^+ also causes vasodilatation (see Chapter 17).

Contractile mechanisms of smooth muscle

No troponin and twice as much actin and tropomyosin are found in smooth muscle compared with striated muscle. Myosin is also present, but at about one-quarter of that found in striated muscle fibres.

Contraction is again caused by an increase in Ca^{2+}. In striated muscle, the contraction is regulated by the interaction of Ca^{2+} with troponin, which regulates actin and myosin cross-bridge function. However in smooth muscle the protein is **calmodulin**, which binds with myosin kinase, activates the myosin heads and causes the cross-bridge mechanism to operate. The energy comes from adenosine triphosphate (ATP) which is degraded to adenosine diphosphate (ADP). The rate at which cross-bridges are formed and released is slower (some 300 times) than that in striated muscle fibres.

A comparison of the properties of skeletal, cardiac and smooth muscle is shown in the Appendix (p. 130).

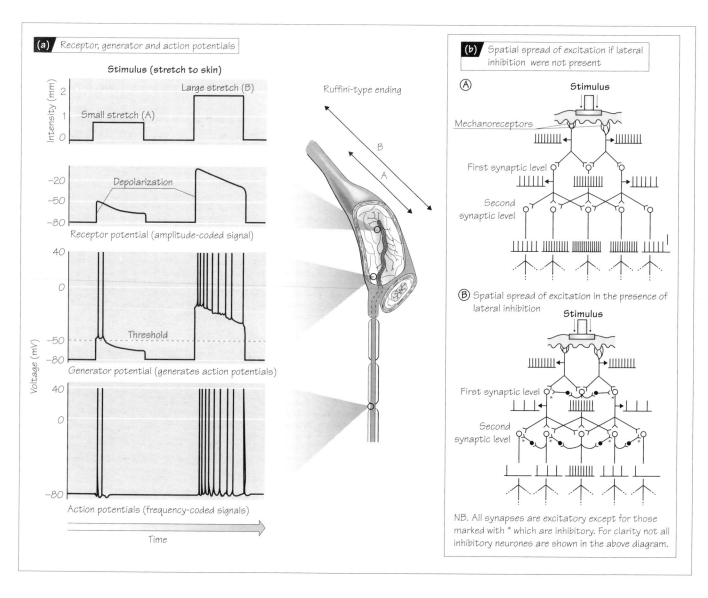

Sensation and perception

The brain obtains its information about the external and internal environment and about the body's relation to the external environment by sensory experience emanating from sensory receptors (sense organs). There are a number of **common steps in sensory reception**: (i) there is a **physical stimulus** (i.e. touch, pressure, heat, cold, light, etc.); (ii) there is a **transduction process** (i.e. the translation of the stimulus into a code of action potentials); and (iii) there is a **response** (i.e. taking a mental note or triggering a motor reaction).

The **specialized nerve ending** (**sensory receptor**), **afferent axon** and its **cell body**, together with the central synaptic connections in the spinal cord or brain stem, are known as **primary afferents**.

The information that can be transmitted to the brain is in the form of action potentials. These action potentials carry this information in the form of **frequency-coded signals** and can signal the following information:

1 The **modality** (**specificity**) of the system. Such modalities include the 'five special senses': sight, hearing, balance, taste and smell. However, it is easy to list others. The skin itself not only senses pressure and touch, but also cold and warmth, vibration and pain (**somatosensation**). In addition, the body senses both the **external environment** and the **internal environment** (its own state). Examples are the sense of equilibrium (**balance**) and a knowledge of the relative positions of the limbs (**proprioception**). Other modalities that are related to information about the state of the body, and that are not directly apparent, are the senses that assess Pco_2 and Po_2, blood pressure, and lung and stomach stretch receptors, the so-called **interoceptors**. Each modality can often be subdivided into further divisions of **quality**, i.e. in the case of taste (sweet, sour, salt, bitter and umami), light (red, green and blue) and hearing (tonal pitches).

2 The **intensity** (**quantity**) of the stimulus (Fig. 54a). The quantity of a sensory impression corresponds to the strength of the stimulus. As the stimulus strength increases, so does the amplitude of the receptor potential (**amplitude-coded signal**) and, when this eventually reaches a **threshold**, it causes action potentials that increase in their frequency of firing as the receptor potential rises (**temporal** or **frequency coding**). Another way in which the strength of the signal is coded is by increasing the number of afferent fibres that are activated (**spatial** or **recruitment coding**).

3 The **duration** of the stimulus. Many receptors will continue to fire impulses as long as the stimulus is applied; others will signal when a stimulus is applied and when a stimulus is removed. However, in most cases, even if a stimulus persists (e.g. constant touch to the skin), the sensation/perception of it wanes. This involves a process called **adaptation**. **Adaptation** occurs at all stages of the transformation of the stimulus: in the transduction process, in the conductance mechanism of the receptor potential, in the synaptic transmission from a secondary sensory cell and in the generation of the action potential. It can also be a function of the central nervous system (CNS) itself once the action potentials reach that far.

4 The **localization** and **resolution** (**acuity**) of the stimulus. The sensory system detects the location of a stimulus, and its fine detail. Both depend on the spacing of receptors (better localization and acuity occur with greater receptor density). The **receptive field** of a sensory neurone itself (sometimes called the **receptor field**) is the area of sensory surface from which that neurone receives an input. Receptor neurones converge onto second-order neurones (usually in the CNS), and then to third- and higher order neurones. These transitions are made in relay nuclei. The receptor field of the primary receptor is usually a small excitatory area. The receptive field of the second-order or higher neurone is larger and more complex (because of both convergence and divergence and excitatory and inhibitory pathways).

The net result is **sensation** and, when interpreted at a conscious level in the light of experience, this becomes **perception**.

Sensory pathways

The coded signals from each of the sensory receptors are relayed to the CNS by peripheral and cranial nerves. Each modality is associated with specific nerves and pathways, e.g. gustatory information is transmitted via facial and glossopharyngeal nerves, and the somatosensory system is transmitted via the dorsal column–medial lemniscal system for the larger afferent fibres (Aα and Aβ) and the anterolateral system (anterior and lateral spinothalamic tracts) for the smaller afferent fibres (Aδ and C).

Each sensory system has its unique pathway into and through the CNS to eventually provide an input into the thalamus. The thalamus, in turn, provides an input to the cortex. Each sensory system projects to a specific area of the primary sensory cortex which is primarily concerned with the analysis of the sensory information, and these neurones, in turn, project to the secondary sensory cortex in which more complex processing occurs. There are further projections to associated areas, such as the posterior parietal, prefrontal and temporal cortices, which can again project to the limbic and motor systems. The latter systems are involved in the processing of the sensory information, leading to responses such as complex behavioural and motor responses.

Lateral inhibition. Figure 54b shows a neural network comprising two mechanoreceptors in the skin and their associated neurones at the next two synaptic levels. The two receptors are each excited equally by a stimulus applied between them. The **divergent** and **convergent connections** seem to impose an avalanche-like spread of excitation at progressively higher levels of the CNS. Pinpoint stimulation appears to lead to an enlarged, less precise and more diffuse representation at each successive synaptic level (**A**). However, this situation is encountered only under pathological conditions (e.g. strychnine poisoning, which blocks inhibiting synapses in the CNS). Inhibition normally prevents the spread of excitation by a phenomenon called **lateral inhibition** (**B**). At each synaptic relay, each excitatory neurone exerts an inhibitory effect by exciting **inhibitory interneurones**. The neurone with the greatest input (the one in the middle) imposes the strongest inhibition on those on either side of it. Lateral inhibition has been shown to exist at all levels of sensory systems: in the dorsal horn of the spinal cord, in the dorsal column nuclei, in the thalamus and in the cortex, as well as in the visual system. The result is an increased spatial sharpening in the CNS of the representation of the distant peripheral stimulus on moving through the synaptic levels.

Descending inhibition. In practically all sensory systems, higher centres can also exert inhibitory effects on all those at lower levels. Such **central inhibition** can act at a point as far peripheral as the receptor or at the afferent ending in the spinal cord. Like lateral inhibition, descending inhibition can be considered to function as a means of **regulating the sensitivity of the afferent transmission channels**.

The types of synaptic mechanism described above indicate that there is great flexibility in the sensory pathways, and that they are not as hard-wired as many pathway diagrams suggest.

55 Sensory receptors

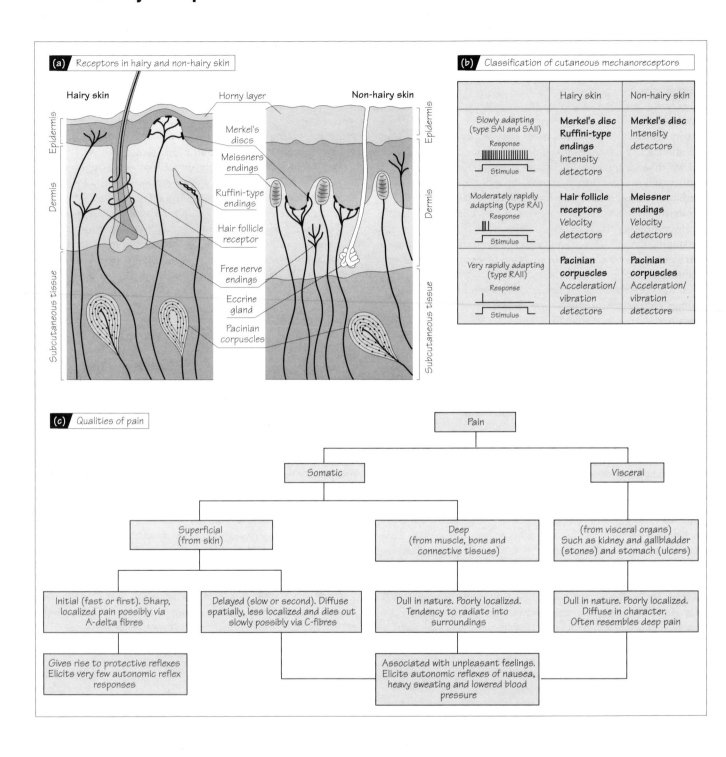

(a) Receptors in hairy and non-hairy skin

Hairy skin

Non-hairy skin

Horny layer

Epidermis

Dermis

Subcutaneous tissue

Merkel's discs
Meissners endings
Ruffini-type endings
Hair follicle receptor
Free nerve endings
Eccrine gland
Pacinian corpuscles

(b) Classification of cutaneous mechanoreceptors

	Hairy skin	Non-hairy skin
Slowly adapting (type SAI and SAII) Response Stimulus	**Merkel's disc** **Ruffini-type endings** Intensity detectors	Merkel's disc Intensity detectors
Moderately rapidly adapting (type RAI) Response Stimulus	**Hair follicle receptors** Velocity detectors	**Meissner endings** Velocity detectors
Very rapidly adapting (type RAII) Response Stimulus	**Pacinian corpuscles** Acceleration/ vibration detectors	**Pacinian corpuscles** Acceleration/ vibration detectors

(c) Qualities of pain

Pain
- Somatic
  - Superficial (from skin)
    - Initial (fast or first). Sharp, localized pain possibly via A-delta fibres
      - Gives rise to protective reflexes Elicits very few autonomic reflex responses
    - Delayed (slow or second). Diffuse spatially, less localized and dies out slowly possibly via C-fibres
  - Deep (from muscle, bone and connective tissues)
    - Dull in nature. Poorly localized. Tendency to radiate into surroundings
      - Associated with unpleasant feelings. Elicits autonomic reflexes of nausea, heavy sweating and lowered blood pressure
- Visceral
  - (from visceral organs) Such as kidney and gallbladder (stones) and stomach (ulcers)
    - Dull in nature. Poorly localized. Diffuse in character. Often resembles deep pain

The **sensory receptor** is a specialized cell. In mammals, receptors fall into the following five groups: **mechanoreceptors**, **thermoreceptors**, **nociceptors**, **chemoreceptors** (see Chapter 56) and **photoreceptors** (see Chapter 57). There is further specialization within these groups. Each receptor responds to one stimulus type; this property is called the **specificity of the receptor**. The stimulus that is effective in eliciting a response is called the **adequate stimulus**.

Transduction processes. Some receptors consist of a nerve fibre alone (e.g. free nerve endings), others consist of a specialized accessory structure (e.g. olfactory receptors, Pacinian corpuscles), and others are more complex and consist of a specialized receptor cell which synapses with a neurone, in other words a secondary sensory cell (e.g. gustatory receptors and Merkel's discs).

Mechanoreceptors. These are found all over the body. Those in the skin have three main qualities: pressure, touch and vibration (or acceleration) (Fig. 55a,b). When the responses to constant stimuli are studied in the various receptors, the receptors can be divided into three types on the basis of their adaptive properties: **slowly adapting receptors** that continue to fire action potentials even when the pressure is maintained for a long period (e.g. **Ruffini's endings, tactile discs, Merkel's discs**); **moderately rapidly adapting receptors** that fire for about 50–500 ms after the onset of the stimulus, even when the pressure is maintained (e.g. **hair follicle receptors, Meissner's corpuscles**); and **very rapidly adapting receptors** that fire only one or two impulses (e.g. **Pacinian corpuscles**) (Fig. 55b). These three types of receptor are examples of receptors in the skin that detect **intensity**, **velocity** and **vibration** (or **acceleration**), respectively.

Free nerve endings. Each skin nerve, in addition to the large myelinated afferents, contains a large number (over 50% of the fibres) of **smaller myelinated and unmyelinated ($A\delta$ and C) axons**. Some of the C fibres are, of course, efferent postganglionic sympathetic fibres. However, a large number of the remaining fibres are afferents that terminate in **free nerve endings** and not in corpuscular structures (Fig. 55a). Many of these are **thermoreceptors** or **nociceptors**.

Thermoreceptors. Thermoreceptors mediate the sensations of **cold** and **warmth**. In the skin of humans, there are **specific cold and warm points** at which only the sensation of cold or warmth can be elicited. These are specific cold and warmth receptors; however, they share the following characteristics: (i) they **maintain discharge** at constant skin temperature, with the discharge rate proportional to the skin temperature (static response); (ii) they have **small receptive fields** ($1\,mm^2$ or less), each afferent fibre supplying only one or a few warm or cold points; and (iii) they serve not only as **sensors for the conscious sensation of temperature**, but also participate (together with temperature sensors in the hypothalamus and spinal cord) in the **thermoregulation** of the body.

Nociceptors and pain. Pain differs from the other sensory modalities with regard to the kind of information it conveys. It informs us of a threat to our bodies when it is activated by noxious (tissue-damaging) stimuli. **Nociception** is defined as the **reception, conduction and central processing of noxious signals**. This term is used to make a clear distinction between these 'objective' neuronal processes and the 'subjective' sensation of **pain**, which is defined as an **unpleasant sensory and emotional experience associated with actual or potential damage, or described in terms of such damage**.

Nociceptors are found in the skin, visceral organs and muscle (cardiac and skeletal) and are associated with blood vessels. The **qualities of pain** are divided into **somatic** and **visceral**. If somatic pain is derived from the skin, it is called **superficial pain**, and, if from muscle, bone joints or connective tissue, it is called **deep pain**. If superficial pain is produced by piercing the skin with a needle, the subject feels a sharp pain; this easily localized sensation fades away rapidly when the needle is removed. This sharp, localized **initial pain** (also called **first** or **fast** pain) is often followed, particularly at high stimulus intensities, by **delayed pain** (also called **second** or **slow** pain), which has a dull (or burning) character with a delay of about 1 s. This delayed pain is more diffuse spatially, dies out slowly and is not so easily localized.

Deep pain (pain from muscles, bones, joints and connective tissues) is dull in nature, poorly localized and has a tendency to radiate into the surroundings.

The responses of the body in terms of both the distress and suffering and the autonomic and motor responses to pain depend on the quality of pain. **Delayed pain** and **deep pain** are accompanied by a **feeling of unpleasantness**, and often elicit autonomic reflexes of **nausea**, **heavy sweating** and **lowered blood pressure**. **Initial pain** gives rise, by contrast, to **protective reflexes**, i.e. flexor withdrawal reflex. **Visceral pain** (pain from organs such as the kidney, stomach and gallbladder) tends to be dull and diffuse in character and resembles deep pain (Fig. 55c).

Histologically, the nociceptors are **free nerve endings** attached to either **$A\delta$ fibres** or **C fibres**. It has been proposed that, in the case of superficial pain, the transmission of **initial (fast)** pain is via **$A\delta$** fibres, whereas **delayed (slow)** pain is signalled by the smaller **C** fibres. The time difference between initial (fast) pain and delayed (slow) pain appears to be explained by the difference in the conduction velocities of the fibres concerned.

Inhibitory influences. Like all other sensory inputs, the nociceptive afferent influx is exposed to inhibitory influences at the receptor, on its way to and through the spinal cord and in the higher levels of the central nervous system. Many of the modern treatments elicit or enhance these inhibitory processes, pharmacologically using drugs, physically using cold or warm wrappings, short-wave radiation, massage and exercise, and by the electrical stimulation of certain structures, including peripheral nerves. **Acupuncture** and **transcutaneous electrical nerve stimulation (TENS)** may possibly depend on the activation and maintenance of inhibitory processes. Naturally occurring **endorphins**, **enkephalins** and **dynorphins** are thought to contribute to these processes. These are endogenous, pain-controlling opiates produced by the body that attach to the specific opiate receptors, so as to inhibit the sensation of pain without affecting the other sensory modalities.

56 Special senses (taste and smell)

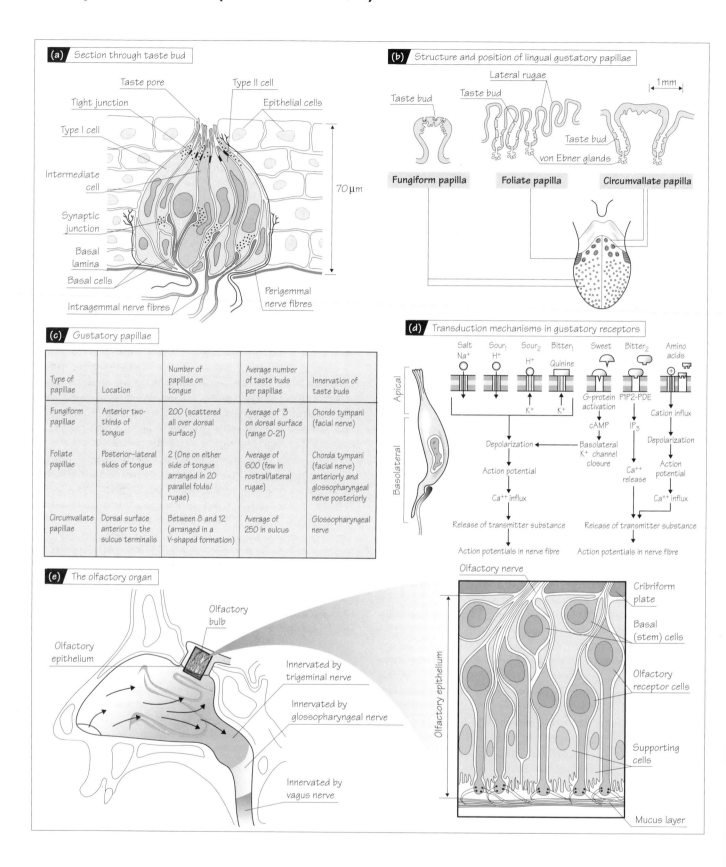

(a) Section through taste bud

Taste pore
Type II cell
Tight junction
Epithelial cells
Type I cell
Intermediate cell
Synaptic junction
Basal lamina
Basal cells
Intragemmal nerve fibres
Perigemmal nerve fibres
70 μm

(b) Structure and position of lingual gustatory papillae

Lateral rugae
Taste bud
Taste bud
1 mm
Taste bud
Taste bud
von Ebner glands

Fungiform papilla   **Foliate papilla**   **Circumvallate papilla**

(c) Gustatory papillae

Type of papillae	Location	Number of papillae on tongue	Average number of taste buds per papillae	Innervation of taste buds
Fungiform papillae	Anterior two-thirds of tongue	200 (scattered all over dorsal surface)	Average of 3 on dorsal surface (range 0–21)	Chords tympani (facial nerve)
Foliate papillae	Posterior–lateral sides of tongue	2 (One on either side of tongue arranged in 20 parallel folds/rugae)	Average of 600 (few in rostral/lateral rugae)	Chorda tympani (facial nerve) anteriorly and glossopharyngeal nerve posteriorly
Circumvallate papillae	Dorsal surface anterior to the sulcus terminalis	Between 8 and 12 (arranged in a V-shaped formation)	Average of 250 in sulcus	Glossopharyngeal nerve

(d) Transduction mechanisms in gustatory receptors

Apical
Basolateral
Salt Na^+
$Sour_1$ H^+
$Sour_2$ H^+
$Bitter_1$ Quinine
Sweet
$Bitter_2$
Amino acids
K^+
K^+
G-protein activation
cAMP
P1P2-PDE
IP_3
Cation influx
Depolarization
Basolateral K^+ channel closure
Depolarization
Action potential
Ca^{++} release
Action potential
Ca^{++} influx
Ca^{++} influx
Release of transmitter substance
Release of transmitter substance
Action potentials in nerve fibre
Action potentials in nerve fibre

(e) The olfactory organ

Olfactory nerve
Cribriform plate
Olfactory bulb
Olfactory epithelium
Basal (stem) cells
Olfactory epithelium
Innervated by trigeminal nerve
Olfactory receptor cells
Innervated by glossopharyngeal nerve
Supporting cells
Innervated by vagus nerve
Mucus layer

The so-called special senses comprise the sensations of **taste**, **smell**, **vision**, **hearing** and **balance**. The receptors involved in taste and smell are **chemoreceptors**, those in vision are **photoreceptors** and those in hearing and balance are **mechanoreceptors**.

The sensations of **taste** and **smell** are two modalities of sense that are very closely related. What the layperson calls 'taste' is really a combination of taste and smell and probably a number of other modalities. Taken together, a better term would be **flavour**. The modalities of flavour are **taste (gustation)**, **smell (olfaction)**, **touch (texture)**, **temperature (thermoreception)** and **common chemical sense (chemoreception)**.

Gustation

Taste buds (the **gustatory end organs**) (Fig. 56a) are found in the tongue, soft palate, pharynx, larynx and epiglottis, and are unevenly distributed around these regions. Those in the tongue are associated with three of the four types of papillae (**fungiform**, **foliate** and **circumvallate**) (Fig. 56b). Those associated with the other oral tissues are found on the smooth epithelial surfaces. The different papillae occupy specific areas of the tongue. Their associated taste buds are innervated by either the **glossopharyngeal (IX) nerve** (posterior one-third of the tongue) or the **chorda tympani branch** of the **facial (VII) nerve** (anterior two-thirds of the tongue). In humans, the number of taste buds varies considerably: on average in the range 2000–5000, but can be as low as 500 or as high as 20 000 (Fig. 56c). Each taste bud is made up of 50–150 **neuroepithelial cells** arranged in a compact, pear-shaped structure (**intragemmal cells**). There is general agreement that there are four types of intragemmal cells: **basal**, **type I (dark cells)**, **intermediate** and **type II (light cells)**. Each taste bud comprises a dynamic system in which there is a rapid turnover of cells within each bud. The **lifespan** of an individual receptor cell is about **10 days**. There is a small opening in the surface, the **taste pore**, where the cells have access to the gustatory stimuli (Fig. 56a).

Since Aristotle (384–322 BC), people have tried to categorize taste into **primary** or **basic qualities of taste**. The four qualities that have stood the test of time are **sweet**, **sour**, **salt** and **bitter**, with a fifth categorized by the taste of monosodium glutamate (**umami**). The mechanisms involved in the process of transduction of the signals that eventually produce these basic sensations of taste are complex. In common with many other receptor cells, gustatory receptor cells use **specifically localized ion channel and receptor sites** for transduction. Unlike many other receptor cells, there is **no single membrane transduction event** and the different basic tastes utilize **different ionic mechanisms** (Fig. 56d). Most gustatory stimuli are water soluble and non-volatile and are either already dissolved or are dissolved in saliva during mastication.

The **common chemical sense** has been defined as the sensation caused by the stimulation of **epithelial** or **mucosal free nerve endings** by chemicals. Evidence suggests that these are **polymodal nociceptors** and that, in the mouth, the major contributor to this sense is the **trigeminal (V) nerve**. The trigeminal innervates almost all regions of the mouth, including the floor of the mouth, the tongue, the hard and soft palate, and the mucosa of the lips and cheek. These nerve endings are stimulated by a number of different chemicals, such as menthol, peppermint, and capsaicin and piperine that are found in chilli peppers and black peppers, respectively.

Olfaction

The human olfactory organ, the **olfactory epithelium** or **mucosa**, is a sheet of cells, 100–200 μm thick, situated high in the back of the nasal cavity and on the thin bony partition (the **central septum**) of the nasal passage. The olfactory system responds to airborne, volatile molecules that gain access to the olfactory epithelium with the in-and-out air flow through and behind the nose. The odour molecules are distributed over the receptor sheet in an irregular pattern by the turbulence of the air flow set up by the turbinate bones (Fig. 56e).

The olfactory epithelium contains specialized, elongated nerve cells (**olfactory receptors**) (Fig. 56e). These cells have very thin fibres that run upwards in bundles through perforations in the skull (the **cribriform plate**) above the roof of the nasal cavity. These bundles of nerves constitute the **olfactory (I) nerve**. They extend only a very short distance, ending in the **olfactory bulbs**, a pair of swellings underneath the frontal lobes. The other end of each olfactory receptor, pointing down into the nasal cavity, is extended into a long process, ending in a knob carrying several hairs (**cilia**) between 20 and 200 μm in length. These cilia are bathed in a thin (35 μm thick) layer of **mucus**, secreted by specialized cells in the olfactory epithelium, in which the molecules of odorous substances dissolve. The molecules diffuse through the surface layer of mucus and stimulate the olfactory receptors. **Hydrophilic** (water-soluble) molecules dissolve readily in the mucus, but the diffusion of less soluble molecules is assisted by 'odour-binding proteins' in the mucus, which are also thought to assist in removing odour molecules from the receptor cells. The mucus layer moves across the surface of the **olfactory mucosa** at 10–60 mm min^{-1} towards the **nasopharynx**. This flow of mucus also assists in the removal of odours after they have been sensed. In the membrane of the cilia are **olfactory receptor proteins**, which interact with the smelly molecules, and initiate a cascade reaction inside the cell that leads to a change in the rate of impulses.

Humans are able to **distinguish 10 000 or more different odours**. Individual olfactory receptor neurones fire off spontaneously at between 3 and 60 impulses per second. When stimulated with particular odours, they increase their firing frequency. Each **receptor cell** responds, but not equally, to many different types of odour.

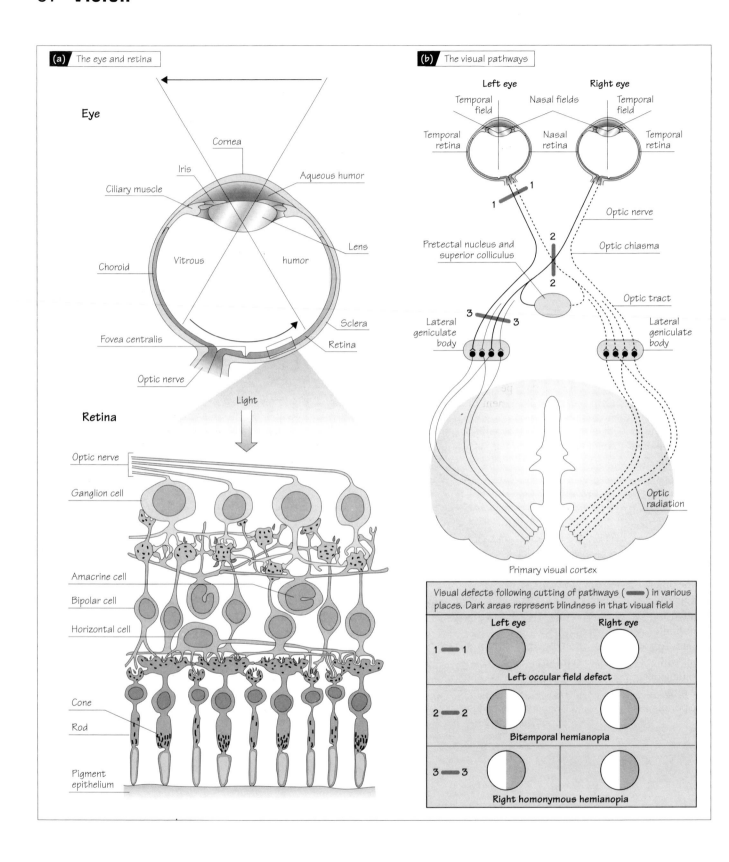

(a) The eye and retina

Eye

Cornea
Iris
Ciliary muscle
Aqueous humor
Lens
humor
Vitrous
Choroid
Sclera
Retina
Fovea centralis
Optic nerve

Light

Retina

Optic nerve
Ganglion cell
Amacrine cell
Bipolar cell
Horizontal cell
Cone
Rod
Pigment epithelium

(b) The visual pathways

Left eye | Right eye
Temporal field | Nasal fields | Temporal field
Temporal retina | Nasal retina | Temporal retina
Optic nerve
Optic chiasma
Pretectal nucleus and superior colliculus
Optic tract
Lateral geniculate body
Lateral geniculate body
Optic radiation
Primary visual cortex

Visual defects following cutting of pathways (▬) in various places. Dark areas represent blindness in that visual field

	Left eye	Right eye
1 ▬ 1	(dark)	(white)
	Left occular field defect	
2 ▬ 2	(half)	(half)
	Bitemporal hemianopia	
3 ▬ 3	(half)	(half)
	Right homonymous hemianopia	

Vision in humans involves the detection of a very narrow band of light ranging from about **400 to 750 nm** in wavelength. The shortest wavelengths are perceived as blue and the longest as red. The eye contains **photoreceptors** that detect light, but, before the light hits the receptors responsible for this detection, it has to be focused onto the retina (200 μm thick) by the cornea and the lens (Fig. 57a).

The photoreceptors can be divided into two distinct types called **rods** and **cones**. **Rods** respond to **dim light** and **cones** respond in **brighter conditions and can distinguish red, green or blue light**. The rods and cones are found in the deepest part of the retina, and light has to travel through a number of cellular layers to reach these photoreceptors. Each photoreceptor contains molecules of the **visual pigments** (rods: **rhodopsin**; cones: **erythrolabe (red)**, **chlorolabe (green)** and **cyanolabe (blue)**); these absorb light and trigger receptor potentials which, unlike other receptor systems, lead to a **hyperpolarization** of the cell and not depolarization.

The layers between the retina surface and the receptor cells contain a number of excitable cells, called **bipolar, horizontal, amacrine** and **ganglion cells**. The **ganglion cells** are the neurones that transmit impulses to the rest of the central nervous system (CNS) via axons in the **optic nerve**. These cells are excited by the vertical **bipolar interneurones** which lie between the receptor cells and the ganglion cells. In addition, this complex structure also contains two groups of interneurones (**horizontal** and **amacrine cells**) that function by exerting their influence in a horizontal manner, by causing lateral inhibition on surrounding synaptic connections between receptor cells and bipolar cells, and bipolar cells and ganglion cells, respectively (Fig. 57a).

Each eye contains approximately **126 million photoreceptors (120 million rods and 6 million cones)** and only **1.5 million ganglion cells**. This means that there is a substantial amount of convergence of receptor and bipolar cells onto ganglion cells, but this is not uniform across the retina. At the periphery, there is a large amount of convergence, but, in the region of greatest visual clarity (the **fovea centralis**), there is a 1 : 1 : 1 connectivity between a single cone receptor cell, a single bipolar cell and a single ganglion cell. The fovea region has a very high density of cones and very few rods, whereas there is a more even distribution of rods and cones in the other regions of the retina.

Each ganglion cell responds to changes in light intensity over a limited area of the retina, rather than to a stationary light stimulus. This limited area is called the **receptive field** of the cell and corresponds to the group of photoreceptors that has synaptic connections with that particular ganglion cell. Ganglion cells are usually spontaneously active. Approximately half of the ganglion cells in the retina respond with a decrease in firing of their impulses when the periphery of their receptive field is stimulated by light, and in-crease their firing rate when the centre of the receptive field is lit up (the **ON-centre cells**); the other half increase their firing rate when the periphery is illuminated and decrease their firing rate when the central receptors are stimulated (the **OFF-centre cells**). This allows the output of the retina to signal the relative brightness and darkness of each area being stimulated within the visual field.

The ganglion cells are further subdivided into two main groups: P cells and M cells. P cells receive the central parts of their receptive fields from one or possibly two (but never all three) types of colour-specific cone, whereas M cells receive inputs from all types of cone. M cells are therefore not colour selective, but sensitive to contrast and movement of images on the retina. The division of P and M cells appears to be maintained throughout the visual pathway and they are involved in visual perception.

The optic nerves from the two eyes join at the base of the skull at a structure called the **optic chiasma** (Fig. 57b). Approximately half of each of the optic nerve fibres cross over to the contralateral side; the other half remain on the ipsilateral side and are joined by axons crossing from the other side. The axons of the ganglion cells from the **temporal region** of the retina of the left eye and the **nasal region** of the retina of the right eye proceed into the left optic tract, whereas the axons from the ganglion cells in the **nasal part** of the left eye and the **temporal part** of the right eye form the right optic tract. The neurones that make up the **optic tract** connect to the first relay stations in the pathway: the **lateral geniculate bodies**, the **superior colliculus** and the **pretectal nucleus** of the **brain stem**. Those fibres that synapse in the superior colliculus and the pretectal nucleus are involved in visual reflexes and orientating responses. A small number of fibres also branch off at this point to synapse in the **suprachiasmatic nucleus**, which is concerned with the body clock and circadian rhythms within the body. However, the bulk of the neurones reach the **lateral geniculate nucleus** in the **thalamus**. Each nucleus contains six cellular layers and the information from the two eyes remains separate, each group of fibres synapsing in three of the layers. The **M ganglion cells** terminate in the lower two layers (called **magnocellular** because the cells are relatively large in these layers). Cells in the magnocellular layers are sensitive to contrast and motion, but not colour. The **P ganglion cells** synapse in the upper four layers of the lateral geniculate nucleus (two for each eye), called the **parvocellular** layers. These layers contain relatively small cells which transmit information about colour and fine detail. The fibres from the lateral geniculate nucleus fan backwards and upwards in a bundle (called the **optic radiation**) through the **parietal** and **temporal lobes** to an area of the cerebral cortex called the **primary visual cortex**. Each cortical cell receives inputs from a limited number of cells in the lateral geniculate nucleus, and therefore has its own receptive field or patch of retina to which it responds.

58 Hearing and balance

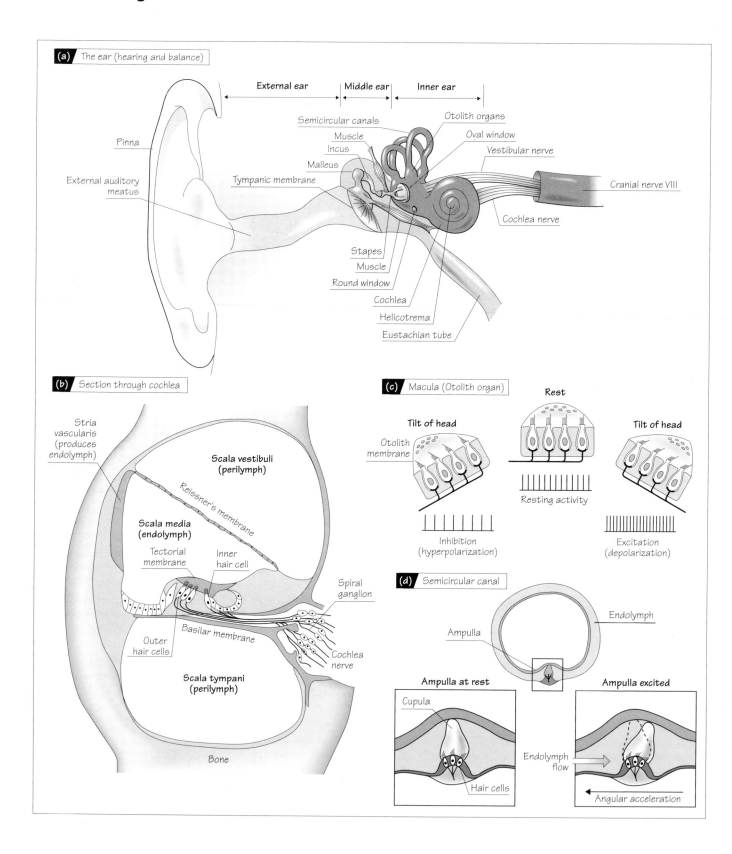

(a) The ear (hearing and balance)

External ear | Middle ear | Inner ear

Pinna
External auditory meatus
Semicircular canals
Muscle
Incus
Malleus
Tympanic membrane
Otolith organs
Oval window
Vestibular nerve
Cranial nerve VIII
Cochlea nerve
Stapes
Muscle
Round window
Cochlea
Helicotrema
Eustachian tube

(b) Section through cochlea

Stria vascularis (produces endolymph)
Scala vestibuli (perilymph)
Reissner's membrane
Scala media (endolymph)
Tectorial membrane
Inner hair cell
Spiral ganglion
Outer hair cells
Basilar membrane
Cochlea nerve
Scala tympani (perilymph)
Bone

(c) Macula (Otolith organ)

Rest
Tilt of head
Otolith membrane
Resting activity
Inhibition (hyperpolarization)
Tilt of head
Excitation (depolarization)

(d) Semicircular canal

Endolymph
Ampulla
Ampulla at rest
Cupula
Hair cells
Ampulla excited
Endolymph flow
Angular acceleration

Hearing

The young healthy human can detect sound wave frequencies of between **40 Hz** and **20 kHz**, although the upper frequency limit declines with age. When sound waves reach the ear, they pass down the **external auditory meatus (the external ear)** to the **tympanic membrane** that vibrates at a frequency and strength determined by the magnitude and pitch of the sound. The vibration of the membrane causes three ear **ossicles (malleus, incus and stapes)** in the **middle ear** (an air-filled cavity) to move, which, in turn, displaces fluid within the **cochlea** (the **inner ear**) as the foot of the stapes moves the **oval window** at the base of the cochlea. This mechanical link prevents the incoming sound energy from being reflected back, and the ossicles improve the efficiency with which the sound energy is transferred from the air to the fluid. Small muscles are attached to the ossicles and contract reflexly in response to loud sounds, thereby dampening the vibration and attenuating the transmission of the sound (Fig. 58a).

The inner ear includes the **cochlea** and also the **vestibular organs** responsible for balance (see later). The receptors involved in both hearing and balance are specialized mechanoreceptors called **hair cells**. Projecting from the apical surface of the hair cell is a bundle of over 100 small hair-like structures called **stereocilia** and a larger stereocilia called the **kinocilium**. Deflection of the stereocilia towards the kinocilium leads to a potential change in the cell (depolarization), the release of a transmitter substance from the base of the hair cell, and activation of the nerve fibres that convey impulses to the higher centres of the brain.

The cochlea comprises a coiled tube of about 3 cm in length (Fig. 58b), with three tubular canals running parallel to one another (**scala vestibuli, scala media and scala tympani**).

The scala vestibuli and the scala tympani contain **perilymph** (which is similar to extracellular fluid in composition), and the scala media contains **endolymph** (similar in composition to intracellular fluid). The scala vestibuli and scala tympani are joined at the tip of the coil (the **helicotrema**); at the base of the scala vestibuli is the **oval window** and at the base of the scala tympani is the **round window**, separating the fluid of the inner ear from the air in the middle ear.

The scala media lies between the two perilymph-filled canals; the boundary between it and the scala vestibuli is called **Reissner's membrane**, and the boundary between it and the scala tympani is called the **basilar membrane**. On top of the basilar membrane sits the **organ of Corti** in which the hair cells are situated. There are around 15 000 hair cells distributed in rows along the basilar membrane. There are two types of hair cell: the **inner hair cells** which form a single row and the more numerous **outer hair cells** arranged in three rows. The hair cells are ideally placed to detect small amounts of movement of the basilar membrane. Because of the changing width of the basilar membrane, high-frequency sounds maximally displace the membrane at the base of the cochlea and low-frequency sounds maximally displace the membrane at the apical end of the cochlea.

The auditory signals are relayed through a complex series of nuclei in the brain stem and the thalamus, eventually reaching the **primary auditory cortex** in the temporal lobe of the cerebral cortex.

Balance

The system associated with balance is called the **vestibular system** and is not only involved with balance, but also postural reflexes and eye movements.

As mentioned earlier, the receptors involved in the vestibular system are hair cells. These hair cells are found in the inner ear in close proximity to the cochlea in two **otolith organs** called the **utricle** and **saccule**, and in a structure called the **ampulla** found in the three semicircular canals. The otolith organs are primarily involved in the detection of **linear motion** and **static head position**, and the semicircular canals in the detection of **rotational movements of the head**.

The four otolith organs (two on each side) each contain a structure called the **macula** which comprises a number of hair cells (Fig. 58c). With the head erect, the macula in each utricle is orientated horizontally and that in each saccule is orientated vertically. The base of each macula contains hair cells whose stereocilia project into a gelatinous mass called the **otolith membrane**. When the head is tilted, the force of gravity displaces the otolith membrane, thereby bending the stereocilia. The nerve fibres innervating the hair cells are spontaneously active: displacement in one direction increases firing and displacement in the opposite direction decreases firing of the neurones. The **utricle** sends signals representing **forwards and backwards movements** and the **saccule** conveys information about **vertical movements**.

The semicircular canals each contain an organ called the **ampulla** (Fig. 58d). They respond to **rotational movement of the head**, and the plane of each canal is perpendicular to the other two, so that, between all six (three on each side), they provide information relating to the rotational acceleration of the head during movement around any axis. Each canal contains endolymph and the ampulla comprises hair cells in which the stereocilia project into a gelatinous mass, with the same specific gravity as the endolymph, called the **cupula**. During acceleration in the plane of a particular canal, the endolymph tends to remain stationary because of inertia. The movement displaces the stereocilia and stimulation of the associated nerve fibres occurs. Again, movement in one direction increases firing of the nerves and movement in the opposite direction causes a decrease in firing. Vestibular afferent fibres from the auditory (VIIIth) nerve have their cell bodies in the **vestibular ganglion** and terminate in one of **four vestibular nuclei in the medulla**. These nuclei also receive inputs from neck muscle receptors and the visual system. They then project to a number of areas of the central nervous system, including the **spinal cord, thalamus, cerebellum** and **oculomotor nuclei**, where they are involved in posture, gait and eye movements. They also project to the **primary somatosensory cortex** and to the **posterior parietal cortex**.

59 Motor control and the cerebellum

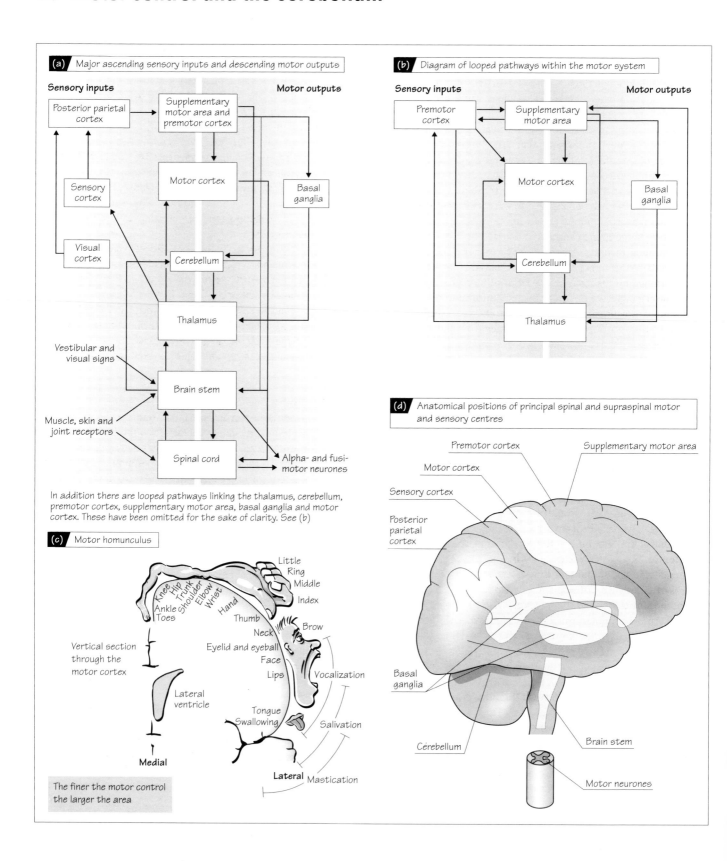

(a) Major ascending sensory inputs and descending motor outputs

Sensory inputs　　　　　　　　　　　　　Motor outputs

Posterior parietal cortex

Supplementary motor area and premotor cortex

Sensory cortex

Motor cortex

Visual cortex

Basal ganglia

Cerebellum

Thalamus

Vestibular and visual signs

Brain stem

Muscle, skin and joint receptors

Spinal cord

Alpha- and fusi-motor neurones

In addition there are looped pathways linking the thalamus, cerebellum, premotor cortex, supplementary motor area, basal ganglia and motor cortex. These have been omitted for the sake of clarity. See (b)

(b) Diagram of looped pathways within the motor system

Sensory inputs　　　　　　　　　　　　　Motor outputs

Premotor cortex

Supplementary motor area

Motor cortex

Basal ganglia

Cerebellum

Thalamus

(c) Motor homunculus

Little
Ring
Middle
Index
Thumb

Knee
Hip
Trunk
Shoulder
Elbow
Wrist
Hand

Ankle
Toes

Neck
Brow
Eyelid and eyeball
Face
Lips
Vocalization

Vertical section through the motor cortex

Lateral ventricle

Medial

Tongue
Swallowing
Salivation

Lateral　Mastication

The finer the motor control the larger the area

(d) Anatomical positions of principal spinal and supraspinal motor and sensory centres

Premotor cortex
Supplementary motor area
Motor cortex
Sensory cortex
Posterior parietal cortex

Basal ganglia

Cerebellum

Brain stem

Motor neurones

Motor control

Motor control is defined as the control of movements by the body. These movements can be both influenced and guided by the many sensory inputs that are received, or can be triggered by sensory events. They can also be triggered by the need to move using internal mechanisms.

The major division of the body into sensory and motor functions is artificial, because almost all motor areas in the central nervous system (CNS) receive sensory inputs.

The organization and physiology of motor systems have been represented as a number of **hierarchical structures**, but these must be viewed with caution, as they are again artificial and, by necessity, oversimplified.

Figure 59a shows the major ascending sensory inputs and descending motor outputs, and Fig. 59b shows the main looped pathways within the CNS.

Voluntary movements can be summarized as follows. Exactly where the idea of a movement is initiated is unknown, but is thought to be in the areas of the cortex other than the primary sensory or primary motor cortices (the **association cortex**) and possibly the **basal ganglia**. At this stage, sensory information relating to the intended movement is analysed in the **posterior parietal cortex**. This sensory information mainly comes from the **visual** and **sensory cortex**.

The posterior parietal cortex activates the **supplementary motor area** and the **premotor cortex**. This excitation also causes the **basal ganglia loop** and the **cerebrocerebellar loop** to be excited and to lead to a degree of amplitude setting and coordination of the activity. The supplementary motor area and the premotor cortex then initiate activity in the **motor cortex**. In addition, the premotor cortex initiates, via the **anterior corticospinal tract** and the connections to the brain stem **ventromedial pathways**, any postural adjustments needed for the movement.

The motor cortex, via the lateral **corticospinal** and **corticorubrospinal tracts**, then initiates the activity of the muscles. This activity is due to the excitation of both α- and **fusi-motor neurones**. During this movement, there is continuous feedback from **receptors** in the **joints**, **muscles** and **skin**, which can lead to fine adjustments via local **spinal** and **brain stem reflexes**. Furthermore, there is often **visual feedback** which can modulate the motor outputs at the cortical and cerebellar levels. Modulations of the activity at all levels continue throughout the voluntary movement.

Figure 59d shows the anatomical sites of the principal motor and sensory centres, and Fig. 59c shows the relative size of the areas in the motor cortex represented by the different parts of the body (the motor homunculus).

The term **upper motor neurones** refers to those neurones that are wholly in the CNS motor pathways. These descending motor pathways are divided into the **pyramidal tracts**, which originate in the cerebral cortex, and the **extrapyramidal tracts**, which originate in the brain stem. The **pyramidal tracts** descend through the **internal capsule** and terminate in the brain stem. One small group of fibres (the **corticobulbar tract**) terminates on cranial motor nuclei and is involved in controlling eye, facial and masticatory muscles. Another larger group of fibres (the **corticospinal tract**) descends directly from the cortex to the grey matter of the spinal cord but, as it passes through the brain stem, it divides into two. Approximately **85%** of the fibres cross over the midline (**decussate**) and descend as the **lateral corticospinal tract**, terminating directly onto the α- and fusi-motor neurones. Some of the fibres do not terminate directly onto the motor neurones but excite interneurones instead. These interneurones can be either excitatory or inhibitory in nature.

The other **15%** of corticospinal neurones, the **anterior corticospinal tract**, do not decussate and remain ipsilateral, eventually terminating in the upper thoracic spinal cord, and project bilaterally onto the motor neurones and interneurones that innervate the muscles of the upper trunk and neck.

The **extrapyramidal tract** neurones project to the spinal cord, where they synapse mainly onto interneurones. There are two groups: the **ventromedial** pathways, which terminate in the motor pools of the axial and proximal limb muscles, and the **dorsolateral** pathways, which terminate in the motor pools of the distal limb muscles.

The **ventromedial pathways** comprise the **vestibulospinal tract**, which receives neurones from the vestibular system and is involved in the reflex control of balance, the **tectospinal tract**, which is involved in the coordination of eye and body movements, and the **reticulospinal tract**, which is concerned with regulating the excitability of extensor muscle reflexes.

The **dorsolateral pathways** comprise mainly the **rubrospinal tract**, which originates in the **red nucleus** in the **midbrain** and projects to similar motor neurone pools as those served by the corticospinal tracts, and are involved with the **reflex control** of **flexor muscles**.

The cerebellum

The cerebellum is anatomically distinct from the rest of the brain and is connected to the brain stem by thick strands of afferent and efferent fibres through **three (cerebellar) peduncles**. Its primary function is the coordination and learning of movements and it is made up of three functional and anatomical structures: the **spinocerebellum**, which is involved in the control of musculature and posture; the **cerebrocerebellum**, which is involved in the coordination and planning of limb movement; and the **vestibulocerebellum**, which is involved with posture and the control of eye movements.

The **spinocerebellum** receives both sensory inputs from the spinal cord and motor inputs from the cerebral cortex. It regulates ongoing movements of axial and distal muscles, by comparison of the descending inputs with the ascending sensory feedback, and regulates muscle tone.

The **cerebrocerebellum** receives inputs from the cerebral cortex, particularly the premotor cortex, and is primarily involved in the planning and initiation of movements, particularly involving the visual system.

The **vestibulocerebellum** receives inputs and sends outputs to the vestibular nuclei in the medulla, and is involved in the regulation of balance, posture and the control of eye movements.

The cerebellum functions by acting as a **comparator**, comparing sensory and motor inputs and achieving coordinated movements that are both smooth and accurate. It can also function as a **timing device** in which it converts descending motor signals into a sequence of coordinated and smooth events. Finally, it can **store motor information** and regularly update it; therefore, given the right sequence of events, it can lead to the initiation of accurate learnt movements.

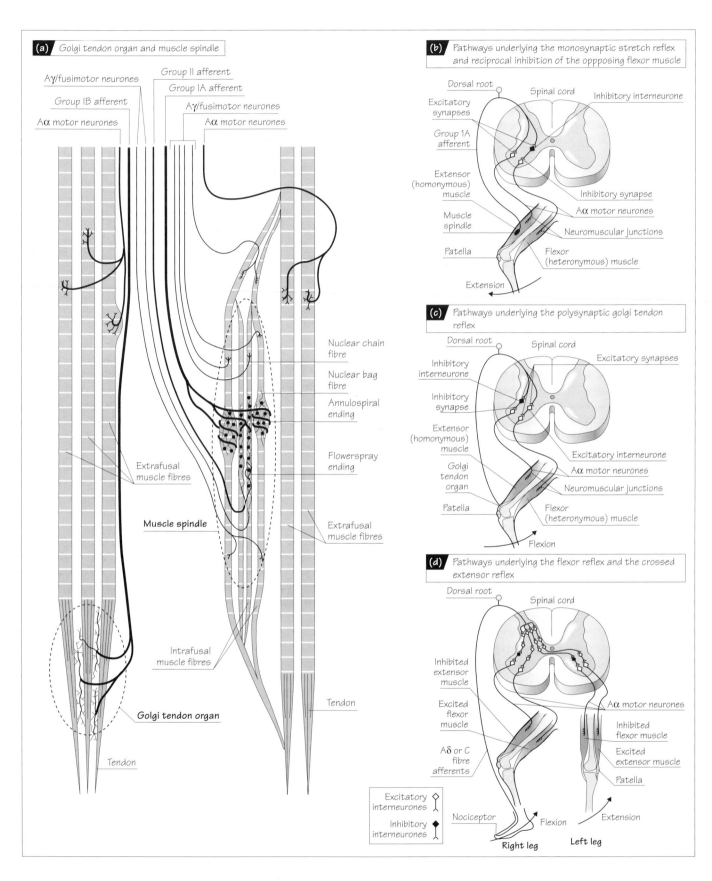

(a) Golgi tendon organ and muscle spindle

Aγ/fusimotor neurones
Group IB afferent
Aα motor neurones
Group II afferent
Group IA afferent
Aγ/fusimotor neurones
Aα motor neurones

Nuclear chain fibre
Nuclear bag fibre
Annulospiral ending
Flowerspray ending

Extrafusal muscle fibres
Muscle spindle
Extrafusal muscle fibres

Intrafusal muscle fibres
Golgi tendon organ
Tendon
Tendon

(b) Pathways underlying the monosynaptic stretch reflex and reciprocal inhibition of the opposing flexor muscle

Dorsal root
Spinal cord
Inhibitory interneurone
Excitatory synapses
Group 1A afferent
Inhibitory synapse
Aα motor neurones
Extensor (homonymous) muscle
Muscle spindle
Neuromuscular junctions
Patella
Flexor (heteronymous) muscle
Extension

(c) Pathways underlying the polysynaptic golgi tendon reflex

Dorsal root
Spinal cord
Excitatory synapses
Inhibitory interneurone
Inhibitory synapse
Excitatory interneurone
Aα motor neurones
Extensor (homonymous) muscle
Golgi tendon organ
Neuromuscular junctions
Patella
Flexor (heteronymous) muscle
Flexion

(d) Pathways underlying the flexor reflex and the crossed extensor reflex

Dorsal root
Spinal cord
Inhibited extensor muscle
Aα motor neurones
Excited flexor muscle
Inhibited flexor muscle
Excited extensor muscle
Aδ or C fibre afferents
Patella
Excitatory interneurones
Inhibitory interneurones
Nociceptor
Flexion
Extension
Right leg
Left leg

We are aware of the orientation of our limbs with respect to one another, we can perceive the movements of our joints and we can accurately assess the amount of resistance (force) that opposes the movements we make. This ability is called **proprioception**. The three qualities of this modality are **position**, **movement** and **force**. The receptors or **proprioceptors** that mediate this modality are principally found in the joint capsules (**joint receptors**), muscles (**muscle spindles**) and tendons (**Golgi tendon organs**).

The **joint capsule** is compressed or stretched when the joint moves, and **mechanoreceptors** within it signal the position of the joint, as well as the direction and velocity of the movement. Individual receptors respond to the position of the joint, as well as the direction and the velocity of the movement, but not the force. The receptor types found in the joint capsule are **Ruffini-type** (slowly adapting) stretch receptors (Chapter 55).

Each muscle contains a number of small muscle fibres (**intrafusal muscle fibres**: 15–30 μm in diameter and 4–7 mm in length) that are thinner and shorter than the ordinary muscle fibre (**extrafusal muscle fibres**: 50–100 μm in diameter and varying in length from a few millimetres to many centimetres). Several intrafusal fibres are grouped together and encased in a connective tissue capsule, called the **muscle spindle**, a specialized receptor that responds to the stretch of a muscle (Fig. 60a). Muscle spindles lie in **parallel** to the extrafusal muscle fibres and are elongated when the muscle is stretched. The primary sensory innervation of the muscle spindle consists of afferent fibres which wind themselves around the centre of the intrafusal muscle fibres (**annulospiral ending**). These are large myelinated fibres (group Ia afferents). These endings are called **primary sensory endings** and, when excited, they evoke a **monosynaptic stretch reflex** involving an excitation of the **homonymous** α-motor neurones and reciprocal inhibition of the **heteronymous** α-motor neurones (Fig. 60b).

Many muscle spindles also have a **secondary sensory innervation** (group II afferent). They are thinner and end in **flower-spray endings** (they are not involved in the monosynaptic stretch reflex).

The intrafusal muscle fibres possess a motor innervation, the Aγ- or fusi-motor neurones. They are smaller in diameter than those innervating the extrafusal muscle fibres, the Aα-motor neurones.

Stretching the muscle, thereby extending both the extrafusal and intrafusal muscle fibres, can excite the muscle spindle. However, there is a second way to excite the primary muscle spindle ending—by contraction of the intrafusal muscle fibre brought about by excitation of the fusi- or γ-motor neurone. This does not change the overall length or tension of the entire muscle, as it is too weak; however, it is sufficient to stretch the central portion of the intrafusal fibres, inducing excitation in the primary sensory ending.

Contraction of the extrafusal fibres can be triggered or at least facilitated by the muscle spindle, either by stretch of the whole muscle or by activation of the fusi- or γ-motor neurones. They can complement each other or have a mutually cancelling effect. The threshold of the stretch reflex can also be varied by intrafusal activity.

The **Golgi tendon organs** are stretch receptors found in the muscle tendons (Fig. 60a). Each receptor is associated with the tendon fascicle of about 10 extrafusal muscle fibres, is surrounded by a capsule of connective tissue and is innervated by large myelinated afferent fibres (group Ib fibres). They are in **series** with the extrafusal muscle fibres and respond to tension in the muscle. They can respond both when the muscle contracts and when the muscle is stretched, unlike the muscle spindle which responds predominantly during stretching of the muscle. Golgi tendon organs protect against overloading and provide a protective reflex. In a functional sense, the segmental connections of the Ib fibres are a mirror image of the Ia fibres. A marked increase in muscle tension, whether resulting from stretch, contraction or a combination of the two, will result in **inhibitory** connections with the **homonymous** motor neurones and **excitatory** connections with the **heteronymous** motor neurones. However, none of these connections is monosynaptic; all involve at least two synapses (Fig. 60c).

The properties of the joint receptors make it very likely that they are primarily responsible for mediating the sense of position and movement. The most likely detectors of force sensation are the muscle spindles and Golgi tendon organs. Other receptors also contribute to the sense of force, as well as movement and position, such as mechanoreceptors in the skin.

Polysynaptic motor reflexes. Many receptors in the body, other than those found in muscle, can trigger motor reflexes. Experiments on animals, in which the spinal cord has been severed, have shown that many of these reflexes are restricted to the spinal cord. They are always polysynaptic. The most prominent examples of these reflexes are the **flexor reflex** and the **crossed extensor reflex** (Fig. 60d). If a pinprick is made to a toe, the stimulated limb is pulled away. There is a flexion of the knee and hip joints. This is called **flexor reflex**. It is a protective reflex, pulling the limb from the site of the noxious stimulus. The delay and the magnitude of the response are very much dependent on the stimulus intensity. The higher the intensity, the shorter the latency and the quicker the response. It can also be observed that the flexion of one limb is always accompanied by the extension of the limb on the other side of the body. In other words, there is an ipsilateral flexor reflex and a contralateral extensor reflex. This contralateral extensor reflex is called the **crossed extensor reflex**. These reflexes are not only enhanced in spinal animals, but also in newborn and premature babies, as during the days just after birth the higher levels of the brain are not fully developed.

Appendix

Comparison of the properties of skeletal, cardiac and smooth muscle

Characteristic	Skeletal muscle	Cardiac muscle	Smooth muscle
Cell shape and size	Long cylindrical cells up to 30 cm long and 100 μm wide	Irregular, branched, rod-shaped cells up to 100 μm long and 20 μm wide	Spindle-shaped cells up to 400 μm long and 10 μm wide
Nuclei	Multinucleated	Mostly single nuclei	Single nuclei
Presence of actin and myosin filaments	Yes	Yes	Yes
Striated (presence of sarcomeres)	Yes	Yes	No
Myogenic activity	No	Yes	Yes
Initiation of contraction	Extrinsic (somatic, neural)	Intrinsic (muscle origin) but influenced by extrinsic autonomic (sympathetic and parasympathetic)	Can be intrinsic via plexus of nerves or extrinsic via autonomic (sympathetic and/or parasympathetic), hormones or stretch
Basic muscle tone	Neural activity	None	Both intrinsic and extrinsic factors
Speed of contraction	Fast	Slow	Very slow
Type of contraction	Phasic	Rhythmic	Tonic with some phasic
Electronic coupling between cells (gap junctions)	No	Yes	A few in multiunit and many in unitary
Influence of hormones on contraction	Small	Large	Large
Effect of nerve stimulation	Excitatory	Excitatory or inhibitory	Excitatory or inhibitory
Spontaneous electrical activity	No	Yes	Unitary (yes), multiunit (no)
Extent of innervation	Each cell innervated	Variable	Unitary (sparse), multiunit (almost every cell)
Site of calcium regulation of contraction	Troponin thin filament	Troponin thin filament	Myosin thick filament
Mechanism of excitation–contraction coupling	Via action potentials and T-system	Via action potentials and T-system	Via action potentials, calcium channels and/or second messengers
Source of activating calcium	Sarcoplasmic reticulum	Sarcoplasmic reticulum and some extracellular	Sarcoplasmic reticulum and some extracellular

Index

Note: page numbers in *italics* refer to figures and tables

sliding filament theory of muscle contraction 41, 109
SMADs 93
small intestine 74, 75, 78, 78–9
    absorption 79
    villi 78, 79
smell sense 120, 121
smooth endoplasmic reticulum Ca²⁺-ATPase (SERCA) 42, 43
smooth muscle 114, 115
    blood vessels 33, 42, 43
    contractile mechanisms 115
    excitation–contraction coupling 42, 43
    innervation 115
    multiunit 115
    properties 130
    visceral 115
sodium ion(s) 15, 19
    channels 18, 19, 20, 21, 39
    concentration gradient 19
    membrane potential 20, 21
    reabsorption 71
    renal proximal tubule 66, 67
sodium channel blockers 71
sodium pump 38, 39, 48, 49, 66, 67, 79
    renal tubular transport 66, 67, 68, 69
    large intestine 83
    thyroid hormones 91
    and action potentials 21
sodium–bicarbonate symporters 73
sodium–calcium exchanger 38, 39
sodium–hydrogen antiporters 73
somatosensation 117
somatostatin 95
somatotrophin see growth hormones
spectrin 17
sperm 103
    capacitation 104, 105
spermatogenesis 100, 101, 103
sphincter of Oddi 80, 81
spinal cord, lateral/ventral horns 24, 25
spinal reflexes 126, 127
spinocerebellum 127
spiral arteries 105
splay 67
spleen 27
staircase effect 39
standard temperature and pressure, dry gas (STPD) 55
Starling's law 37, 40, 41, 46, 47
    glomerular filtration rate 64, 65
    renal tubular transport 67
stereocilia 125
stimulus intensity 117
stomach 74, 75, 76, 77
    gastric emptying 78
    mechanoreceptors 76, 77, 117
stress 98, 99
    milk ejection reflex inhibition 107
stroke volume 33, 37, 40, 41
submucosal glands 50, 51
submucosal plexus 74, 75
substance P 42, 43
suckling 107
supplementary motor area 126, 127
surface tension, alveoli 53
surfactant 51, 52, 53
swallowing 75, 77
swallowing centres 75, 77
sweat glands 24, 25
sweating 49
sympathetic chains 24, 25
sympathetic cholinergic neurones 25
sympathetic nervous system 24, 25, 41
    adrenal medulla 99
symporter 19, 67, 73
synapse 25
syncope 45
syncytium, functional 115

systemic circulation 32, 33
systole 33, 35
    atrial/ventricular 37
systolic pressure 32, 33

T cells 28, 29
T helper cells 28, 29
T wave 37
taeniae coli 82, 83
taste
    qualities 121
    sense 120, 121
taste buds 120, 121
taste pore 120, 121
teeth 75
temporomandibular joint (TMJ) 75
tendons 108, 109
TENS (transcutaneous electrical nerve stimulation) 119
testosterone 100, 101, 103
    binding proteins 84
    secretion 85
tetanus 112, 113
tetany, hypocalcaemic 97
thalamus 126
thalassaemia 27
Thebesian veins 35
thermogenesis 91
thermoreception 121
thermoreceptors 119
thermoregulation 48, 49, 119
thiamine 83
thirst 70, 71
thoracic cage 50, 51
thoracic duct 79
threshold potential 20, 21
thrombin 26, 27
thromboplastin 26, 27
thrombosis 43
thromboxane A₂ 26, 27, 42, 43, 49
thrombus formation 26, 27
thymine 10, 11
thymus 29
thyroglobulin 90, 91
thyroid gland 85, 90, 91
thyroid hormone receptors 84, 91
thyroid hormones 84, 90, 91, 95
    calcium effects 97
thyroid-responsive element (TRE) 91
thyroid-stimulating hormone (TSH) 85, 90, 91
thyroxine 85, 90, 91
tidal volume 50, 51
tight junctions 42, 47
    large intestine 83
    renal tubules 62, 63
tongue 74, 75, 120, 121
tonic depression 45
total lung capacity 50, 51
total peripheral resistance 44, 45, 49
transcapillary exchange 47
transcellular fluid 15
transcription, genetic 10, 11
transfer factor 55
transforming growth factor b (TGF-b) 95
translation, genetic 10, 11
transmembrane proteins 17
transport, primary/secondary active 19
transporter proteins 17, 66, 67
transudation 53
Treppe effect 39
tricuspid valve 34, 35, 37
triglycerides 87
triiodothyronine (T₃) 85, 90, 91
tRNA 10, 11
trophoblasts 104, 105
tropomyosin 108, 109
troponin 41, 108, 109
T-tubules 38, 39, 109
tubular transport maximum 66, 67

tumour necrosis factor (TNF) 29
tympanic membrane 124, 125
tyramine 25
tyrosine kinase 84

ultrafiltration 64, 65
uncoupling proteins (UCPs) 91
uniporter 19, 67, 68, 69
uptake-1 and -2 25
uracil 10, 11
urea 68, 69
    renal proximal tubules 67
urethra 63
urethral sphincter 63
urine
    hypotonic 71
    osmolality 70, 71
    ouput 63
utricle 125

v waves 37
vagus nerve 75, 76, 77, 80, 81, 83
vasa recta 62, 63, 69
vasa vasorum 43
vasoactive intestinal polypeptide (VIP) 85
vasoconstriction 42, 43, 48, 49
    angiotensin II 71
    antidiuretic hormone 70, 71
    hypoxic pulmonary 49
vasodilatation 42, 43, 49
vasopressin see antidiuretic hormone (ADH)
veins 32, 33
    valves 33
vena cava 32, 33
venous plexus 49
ventilation 58, 59
ventilation–perfusion matching 49, 60, 61
ventilation–perfusion mismatch 60, 61
ventricles 32, 33, 34, 35
ventricular ejection 36
ventricular filling 36
ventricular function curve 41
ventricular muscle, action potentials 39
ventricular pressure–volume loop 36, 37
ventromedial pathways 127
venules 32, 33, 46, 47
verapamil 43
vesicles 24, 25
vestibular ganglion 125
vestibular organs 124, 125
vestibular system 125
vestibulocerebellum 127
viscosity 30, 31
vision 121, 122, 123
visual cortex 126, 127
    primary 122, 123
visual feedback 127
visual pigments 123
vital capacity 50, 51
vitamin B12 27, 79, 83
vitamin D 63, 68, 69, 96, 97
vitamin K 27, 83
vitamins 79
voltage gating 18, 19

water
    reabsorption 70, 71
    renal proximal tubule 66, 67
water compartments of body 14, 15
white blood cells (leukocytes) 27

X chromosomes 103

Y chromosomes 103

zona pellucida 104, 105
ZP3 104, 105
zygote 104, 105